DISSOCIATION, ENZYME KINETICS, BIOENERGETICS

SECOND EDITION

HALVOR N. CHRISTENSEN, Ph.D.

Professor, Department of Biological Chemistry
The University of Michigan
Ann Arbor, Michigan

A LEARNING PROGRAM
FOR STUDENTS OF
THE BIOLOGICAL
AND MEDICAL SCIENCES

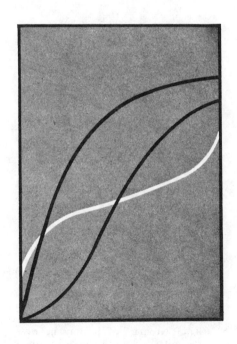

Dissociation, Enzyme Kinetics, Bioenergetics   ISBN   0-7216-2580-0

PREFACE _____

The scope of these three learning programs, combined into one book, is suitable for many introductory biochemistry courses. The author's three earlier separate programs, which are somewhat longer and more detailed, will probably be preferred for students specializing in biochemistry, biophysics and the like. I will not speculate, beyond that prediction, on the line of preference between the two levels of presentation now offered. Depending on personal objectives, the instructor or student may select from the following list of subjects what is to be learned from the study of the present program:

I.   pH and Dissociation
1. conversion of selected values of $[H^+]$ to pH, and vice versa; calculations of pH with the Henderson–Hasselbalch equation
2. the meaning of titration curves; comparison of the amounts of H^+ required to change the pH of water and the pH of solutions of weak bases. (Corrections for the portions of titration curves that fall below pH 3 and above pH 11 are not considered.)
3. selection of indicators for acid-base titrations and for pH measurements
4. buffering and its distribution on the pH scale
5. titration of carbonic and phosphoric acid
6. titration of NH_4^+ and NH_3; conjugate acids and bases

III.   Useful energy from metabolic reactions
1.   that living cells have only one general way to get energy from chemical reactions
2.   that this usable or *free* energy can be measured by reference to the equilibrium constant, because its amount depends on the strength of the tendency of energy-yielding reactions to go toward equilibrium
3.   that no energy-requiring process can occur in the cell unless it is made a part of an energy-releasing process, and that enzyme systems produce coupling by opening a pathway around the barrier of an endergonic step
4.   the necessity of chemi-osmotic coupling, and its bidirectional operation
5.   that the potentials against which electrons can be driven in oxidation-reduction reactions can also measure the free energy changes of these reactions, and therefore predict which reactions will drive which reactions when successfully coupled
6.   the problem of the transduction of the free energy of respiratory metabolism to oxidative phosphorylation as an important example of energetic coupling
7.   the terms *free energy, endergonic, exergonic, spontaneous reaction, thermodynamically reversible process, energetic coupling, active transport, activation energy, half-cell reaction, standard electrode potential, standard half-cell potential.*

Once the three major subjects are combined into one volume, it becomes easier to develop instructive interactions among them. Some of these are introduced in the texts of the three parts, and others are suggested by questions in the appendix. Examples are the plotting of dissociation behavior not involving the hydrogen ion, allosteric phenomena in the association of O_2 to

the hemoglobin molecule and in enzyme regulation, thermodynamic limitations to regulatory effects on enzyme action, and also
kinetic and thermodynamic approaches to membrane transport.

An important consideration for the optimal use of learning
programs is that they be inexpensive enough so that the student
does not hesitate to write his answers into them at each perusal.
Learning is not nearly as effective if economy constrains the
student merely to vocalize or imagine his answer. In an effort to
approach such an ideal, the appendices included here are limited
to short sets of test problems and their answers. Additional test
problems and other supporting material available for assignment in
the longer programs should accordingly be accessible to the
student in the form of library copies. Even the textual part of
those programs may be used as a reference by the user of the
present work, depending on his personal interest.

In the classroom testing of the present work another benefit
of using it in combination with the longer programs has been
observed: After a student has perused a learning program two or
three times, he or she benefits from the modified and therefore
somewhat unexpected sequence presented by an alternative form
of the same program. The instructor should notice, however, that
titration curves are presented here with the pH on the ordinate to
enhance possible coordination of its use with widely used
textbooks, whereas in our more comprehensive program the pH is
placed on the abscissa.

Biological aspects of pH and of kinetics are treated quite as
fully in the shorter versions as in the longer; introductory and
summary sections are also proportionately larger. Furthermore, a
new, short biological section, distantly related to a fourth learning
program, "Neutrality control in the living organism," has been
added to the section on pH and Dissociation under three headings:
The bicarbonate-carbonic acid buffer system under pulmonary
control; Balancing metabolic H^+ release and fixation in the
organism; Ultimate correction of the neutrality by the kidney. I
believe this short section touches on some of the best ideas that

biochemistry has to offer to the important area of the application of dissociation behavior to biological function in both its physiological and pathological aspects.

Part III, a new, short program called Useful Energy from Metabolic Reactions, cannot reasonably be described as a contraction of the longer learning program, *Introduction to Bioenergetics: Thermodynamics for the Biologist*, by myself and Richard A. Cellarius of Evergreen State College, Olympia, Washington. Although both of these bioenergetics programs presuppose essentially no training in thermodynamics, they differ in that the first, which is about five times as long, addresses itself to the biology student who stands to benefit from a number of aspects of that subject, whereas the present short program focuses on the aspects of energy transfer most needed in introductory courses in biochemistry. The resulting, somewhat restricted, emphasis may be seen by considering the three principal headings treated here: Definition of free energy change; The meaning of coupling; Oxidation-reduction reactions and potentials (verging into oxidative phosphorylation). What will not be evident from my listing of these three headings is the use here of membrane transport to sharpen the definition of energetic coupling between chemical reactions — a device which incidentally places the subject of active transport as a subheading under bioenergetics.

Biochemists and other biologists are presently grappling with the problem of how to integrate into their courses the newer molecular knowledge of membrane function. A partial and provisional solution to that problem is offered here, although a further consideration of transport in those courses, outside the scope of these programs, under the heading of integration of metabolism (perhaps along with a brief definition of mediation of transport under kinetics, as illustrated here in Part II, Items 29 to 31), seems to me desirable to any complete and modern biochemistry course. Besides helping you find the best place to teach transport in your courses, my hope is to use transport here to help identify the essence of bioenergetic coupling.

From reading what biologists write about the energization of metabolic processes, particularly of transport, I am convinced that the biochemist is not always successful in teaching the meaning of energetic coupling. One can search out the defect from some of our principal textbooks. During the short, early interval in our courses when we discuss how the free energy made available from one reaction is harvested in the progress of another, we may assert that energetic coupling arises solely from the participation of a common intermediate in the two sequential reactions. Otherwise, true to the spirit of thermodynamics, we tend not to provide biochemical examples of mechanism at that point.

Later in the course, when we come to explain how an ester or an amide link can be generated by coupling the process to the breakdown of ATP, we tend to use the term energetic coupling in a sense subtly different from the one defined earlier. We see ATP synthesized without ADP + P_i being brought to high concentrations by a preceding reaction; we see a dipeptide synthesized even though its component amino acids remain at low levels. Instead, the endergonic reaction, at this level in the course, has been mechanistically fused to another reaction sufficiently exergonic to leave the whole coupled process spontaneous. If the student reminds us of our first definition of coupling, we can inevitably show him a sequential linkage in the new pathway — but he may be more impressed with our legerdemain than with the clarity he reaches on the matter.

This bioenergetics section represents an area for which teaching is in a much less developed and standardized state than in the other areas. Perhaps I may say that it stands now in the same, unrecognized need of development that the teaching of hydrogen-ion dissociation presented some 15 years ago. Hence the present codification of the subject will be seen as having novel aspects, which I hope may prove stimulating. The instructor will need to give his students the chance to communicate to him how well this new codification may enhance comprehension of bioenergetics by the nascent mind. As for all instructional material of external origin, the instructor should make himself aware of any needed contextual instruction.

As always, the author welcomes suggestions as to how the treatment of these three areas may be improved or made more suitable for the students to whom it is directed.

HALVOR N. CHRISTENSEN

Comment on the Second Edition:

The economics of publication require that the Second Edition be produced on a smaller scale and distributed under supervision of the author if its price is to remain tolerable. Revision has aimed also at the latter goal: A few errors have been corrected, and a few pages omitted with the loss of four little-used items from Part III. A sixth, easier Bioenergetics problem has been added on page 182. The author hopes by a similar conservative course to keep in print other of his learning programs while colleagues continue to insist on their utility.

CONTENTS _____

CONTENTS

pH AND DISSOCIATION

DEFINITION OF pH

1

The subject of acid-base balance is concerned with relationship between weak acids and the hydrogen ion. Because we deal mainly with very low concentrations of the _H^+_ , we must consider the special scale used for recording these concentrations, namely the pH scale.

1

hydrogen ion

2

Rather than deal with awkward values like $0.00001N$ or $1 \times 10^{-5}N$ for the hydrogen ion concentration, we follow the suggestion of Sørensen and using the exponent only — without its minus sign — we say that the _pH_ is 5.

2

pH

3

In other words, if the hydrogen _ion concentration_ (which we shall write as [H^+]) is $10^{-6}N$, we assign a pH of _6_ .

3

ion concentration

6

4

Sørensen's _pH_ scale permits us to represent the enormous range of [H^+] from a one normal to a $0.00000000000001N$ [H^+] with numbers extending from _zero_ to 14; or if we have need, we can go even beyond this range.

4

pH

zero

5

You will recall that the term *logarithm* expresses the power to which 10 must be raised to represent a number. Therefore the values *5* and *6* given in

Items 2 and 3 are the *negative* _____ of the [H$^+$] concentrations, 10^{-5} and _____ N.

6

Hence the _pH_ is the negative logarithm of the _hydrogen ion concentration_ ; that is, pH = -log [H$^+$].*

7

Since log x - log y = log $\frac{x}{y}$, and -log y = log $\frac{1}{y}$, the pH may be defined either as -log [H$^+$] or as log _____ .

8

Suppose that the hydrogen ion concentration is 0.00002N, that is, 2×10^{-5}_____. If you are familiar with logarithms, you can use a log table to determine that the -log of 2×10^{-5}, and hence the pH, is _____. (If this is evident to you, proceed directly to Item 16. If Items 7 and 8 are *totally* unclear to you, you should refer to a textbook or learning program on the use of logarithms.)

9

To obtain this value, you use your log table to determine that the logarithm of 2 is _____ ; whereas we know by definition that the logarithm of 10^{-5} is _____.

5

logarithms

10^{-6}

6

pH

hydrogen ion concentration

7

$\frac{1}{[H^+]}$

8

N

4.7

---

*This statement involves an approximation that may be known to you. Actual measurements of pH are made by an electrode sensitive to the *activity* and not the *concentration* of hydrogen ion. Indeed, it is not quite correct to say that pH = -log a$_{H^+}$. We should make a serious mistake, however, to let our recognition that this relation is approximate dissuade us from practicing the translation of pH into [H$^+$] and *vice versa*, because it is of the greatest importance that we attain a sense of the nature of this mathematical relation.

9

0.30

-5

10

negative

11

-5.00

4.70

12

0.60

logarithm

13

-7.40

7.40

10

Hence for the log of 2×10^{-5} we need to add -5 and $+0.30$. Sometimes one will see such a sum written as a partially negative, partially positive logarithm, like this: $\bar{5}.30$. But to represent the pH properly, we need an entirely _____ logarithm.

11

A value of -4.70 corresponds to the sum of _____ and $+0.30$. Therefore, a $[H^+]$ of 2×10^{-5} is represented by a pH of _____ .

12

Using your log table to determine the pH corresponding to $[H^+] = 4.00 \times 10^{-8} N$, you will see that the logarithm of 4.00 is _____. We can, if we like, write the _____ for 4×10^{-8} as partially negative and partially positive, thus: $\bar{8}.60$.

13

If we then make this logarithm entirely negative, we obtain the result _____. Hence the pH is _____ .

14

Some students may prefer to look at this operation in a slightly different way. The logarithm of a product is always the sum of the logarithms of the numbers to be multiplied. Hence,

$-\log [H^+] = -\log (2 \times 10^{-5})$ becomes

$-\log [H^+] = -(\log 2 + \log 10^{-5})$

$\qquad\qquad = -\log 2 - \log 10^{-5}$

By the laws of logarithms,

$$\log 10^{-5} = -5 \log 10$$

Hence,

$$-\log [H^+] = -\log 2 + 5 \log 10$$

$$= \underline{\hspace{1.5cm}} + \underline{\hspace{1.5cm}} = \underline{\hspace{1.5cm}} .$$

The pH is _____ .

15

Notice above that in order to indicate that the measurements and calculations have been carried out to a certain degree of accuracy we retained two zeros to the right of the decimal point in our value for the $[H^+]$, also _____ decimal places to the right of the decimal point in the pH. It is an important habit to retain such digits, even when they are zeros, when they have significance, and to _____ them when they do not.

16

In exchange for the convenience of using pH values from 0 to 14 to represent a very wide range of H+ concentrations, we pay a price. We may not readily sense that a pH of 4.7 stands for twice the hydrogen _____ represented by a pH of 5, or that a pH of 6 represents one-tenth the $[H^+]$ represented by a pH of _____ . If this matter is evident to you, proceed directly to Item 22. Otherwise, let us proceed to clarify these relationships.

17

Note that when the pH rises from 5 to 6, the $[H^+]$ falls from 10^{-5} to 10^{-6} or to _____ of its initial value. Note also that when the pH falls from

---

14

$-0.30 + 5.00 = 4.70$

4.70

15

two

drop (omit)

16

ion concentration

5

5 to 4.7, the $[H^+]$ rises from 1×10^{-5} to $2 \times \underline{\hspace{1cm}} N.$

17

one-tenth

10^{-5}

18

doubled

rises

ten

19

$[H^+]$

8.00×10^{-8}

20

$[H^+]$

10^{-7}

18

A rise of one unit in the pH accordingly means that $[H^+]$ has been decreased to one-tenth. A fall of 0.3 units means that the $[H^+]$ has been $\underline{\hspace{2cm}}$. These relationships follow from the fact that 1 is the logarithm of 10, and 0.3 the logarithm of 2. As the pH falls from 7.4 to 6.4, the $[H^+]$ $\underline{\hspace{1cm}}$ to $\underline{\hspace{1cm}}$ times its initial value.

19

Suppose that we have verified that a pH of 7.40 represents a $[H^+]$ of $4.00 \times 10^{-8} N.$ If the pH is decreased from 7.40 to 7.10, we may quickly sense without consulting a log table that the $\underline{\hspace{1cm}}$ has been doubled to $\underline{\hspace{1cm}} N.$

20

Similarly a pH of 6.70 corresponds to a value for the $\underline{\hspace{1cm}}$ equal to $10^{-6.70}$, or $10^{7.30}$. The log table shows that the logarithm 0.30 corresponds to the number 2.00. Hence, the $[H^+] = 2.00 \times \underline{\hspace{1cm}} N.$

21

But more quickly, by our rule of thumb we might have sensed that a pH of 6.70 would represent $\boxed{\text{one half/twice}}$ the $[H^+]$ represented by a pH of 7.00.

(Cross out the inapplicable answer; the correct response appears in the answer column.)

twice

THE TECHNIQUE OF TITRATION

22

We use the technique of *titration* to discover the presence of substances in solution that bind H^+. When we titrate we purposely vary the $[H^+]$ over the range in which we are interested, to see what _____-binding or _____-releasing substances we encounter.

23

It is true that the hydrogen ion is hydrated in water solution, and that a considerable part of it is present as the hydronium ion, H_3^+O. If our intention were to compare the behavior of H^+ in aqueous and nonaqueous solutions, we should need to take the formation of the _____ into account. Since in biology and medicine we deal almost exclusively with | aqueous/nonaqueous | solutions, we can, for our present purposes, neglect the hydration of the hydrogen ion. We would achieve neither rigor nor consistency by showing it always as the monohydrate.

24

In this technique of titration, we will add HCl or NaOH solution to cause the pH to vary. In using these reagents we come very close to adding pure

---

22

H^+

H^+

23

hydronium ion (H_3^+O)

aqueous

H^+ and OH^-, because these reagents probably are entirely dissociated in water. The other ions they contain, _____ and _____, are inert and do not enter into reactions with H^+ and OH^- in water solution.

24

$\left.\begin{array}{l} Cl^- \\ \\ Na^+ \end{array}\right\}$ (either order)

25

H^+

titration

26

milliequivalents

25

Before we try to recognize _____ binding agents by the technique of _____ we should consider how the pure solvent behaves, when H^+ or OH^- is added.

26

Suppose that we take 50 ml of pure water. The initial pH may be about 6 or 7, depending on how pure the water really is, and on the presence or absence of dissolved CO_2. Let us begin to titrate by adding a drop (say 0.05 ml) of N HCl. With a glass electrode we will find that the pH has fallen to 3. Normal HCl contains one milliequivalent of H^+ per milliliter (ml); hence, adding 0.05 ml of it introduces 0.05 _____ of HCl into a total volume of 50.05 ml of solution.

27

Notice in this connection that the unit of normality is equivalents per _____litre_____ or _____mE_____ per milliliter. Therefore, if we multiply the normality of a reagent by the number of ml of the reagent taken, our answer will be expressed in (mE/ml) $\times$ ml, or more simply, in _____ .

28

The solution resulting from the addition of 0.05 ml N HCl to 50 ml water contains 0.001 milli-equivalents H$^+$ per ml, or 0.001 equivalents per _____ ; hence _____ = $10^{-3}N$, and pH = _____ .

29

In the same way we can find experimentally that 1 drop of N NaOH brings the _____ to 11. We will not stop now to calculate why this value is reached.

30

Unless we are titrating an unusually dilute solution, we can regard additions of less than 0.05 mE of the reagents HCl or NaOH as negligible in relationship to the total amounts required in the titration. Therefore, in the range between pH 3 and 11, we may neglect the quantities of these reagents required to vary the pH of pure water. (Below pH _____ and above pH _____ this is no longer the case. For example, to establish a pH of 1 would require 0.5 ml/5 ml of N HCl for a final volume of 50 ml.)

31

We will, however, deal mainly with the pH range, 3 to 11. Suppose that instead of water we take for our titration acetic acid, 50 ml, 0.1 N in concentration. The initial pH is about 3. As we proceed to add NaOH, we find that for this solution far more is needed to _____ the pH.

27

liter

milliequivalents ⎫ (either
                  ⎬
mE   ⎭ form)

28

liter

[H$^+$]

3

29

pH

30

3

11

5 ml

THE TECHNIQUE OF TITRATION

raise (change)

32

Here we see coordinate axes labeled *pH* on the ordinate, and *ml alkali added* on the abscissa. Mark points on the graph to show the following results, which were observed by measuring the pH with a glass electrode after each addition of _____ .

ml N NaOH	pH
0.0	3.00
0.5	3.75
1.0	4.10
2.0	4.52
3.0	4.88
4.0	5.30
4.5	5.65
4.95	6.70
5.00	9.0
5.05	11.0

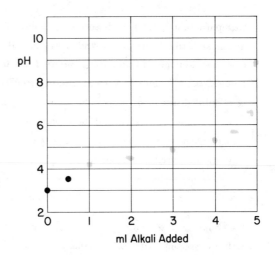

33

In the above titration the addition of OH^- has lowered $[H^+]$ as follows: $H^+ + OH^- \rightarrow H_2O$, causing more acetic acid to dissociate: $HA \rightarrow A^- +$ _____. We may represent the overall action of the NaOH as the sum of these two reactions, as follows: $HA + OH^- \rightarrow$ _____ + _____ .

34

Note that we observed the pH at several points during this titration. In the past, you probably have titrated acetic acid with NaOH in the presence of an indicator. In that case a color change of this _____ told you when a given _____, perhaps 8.5, was reached or passed. During the other stages of the titration you | were/were not | aware of the pH.

35

Our purpose now is a different one, namely to understand the behavior of weak acids and the hydrogen ion; and for this purpose we need to know how the _____ changes at every step as acid or alkali is added. To put the matter another way, we need to understand how various substances react with respect to _____ as its concentration is raised and lowered.

36

We could trace the above titration curve quite as well by starting with $0.1N$ sodium acetate and titrating with N HCl. On our graph we would then

32

NaOH

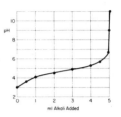

ml Alkali Added

33

H^+

$A^- + H_2O$

(either order)

34

indicator

pH

were not

35

pH

H^+

plot *acid added* toward the left. In this case we increase [H$^+$], causing the acetate to bind H$^+$ as follows: A$^-$ + _____ → _____ .

36

H$^+$, HA

37

3

38

acetate

acetic acid

is somewhat

37

Probably you never have titrated acetate in the laboratory with HCl; but, given an appropriate indicator showing a color change just below pH 3/8.5 , one could estimate the concentration of an acetate solution by titrating it with HCl.

38

Consider that when we study the nature of acetic acid by titrating it, we may either convert acetic acid to _____ , or convert acetate to _____ to discover in what pH range this interconversion occurs. From this situation we see that the term *neutralization* is somewhat/is not ambiguous when applied to these titrations.

39

The term neutralization therefore probably should/should not be understood to mean the bringing of a solution to neutrality (in the sense of pH 7.0). It can be applied equally well for the conversion of acetic acid to _____ , or the conversion of acetate to _____ . Therefore this term must be used with caution.

40

Notice that the titration of acetic acid is substantially complete at pH 8. Above pH 9, our acetic acid solution behaves essentially ｜like/unlike｜ pure water on titration.

CONJUGATE ACIDS AND CONJUGATE BASES

41

The same would be found to be true below pH 2. This is an important point: *A weak acid is a substance that releases H^+, but its behavior in this regard* ｜occurs at all pH values/is limited largely to a particular range of $[H^+]$｜ .

42

Brønsted suggested that we use the term *base* for a substance that can *bind* H^+. Again, this will occur mainly in a particular range of _____ concentration, and one sufficiently high to cause a given weak _____ to accept H^+.

43

In the case of the titration we have just studied, acetic acid and the acetate ion are called a

39

should not

acetate

acetic acid

40

like

41

is limited largely to a particular range of $[H^+]$

42

H^+

base

conjugate pair: a weak acid and its *conjugate*

_____ ; or a weak base and its

_____ .

43

base

conjugate acid

ORIGIN OF THE FORM OF THE TITRATION CURVE; THE HENDERSON-HASSELBALCH EQUATION

44

To understand why the pH rises exactly as it does along a sigmoid curve, we must turn to the mass–action law. The mass–action law states that for the reaction, $HA \rightleftharpoons H^+ + A^-$, at equilibrium the concentration of the three substances involved will have the relationship,

$$\frac{[H^+] \times [A^-]}{[\text{_____}]} = K'$$

(Because we will find it convenient to use the *uncorrected concentrations* of HA and A^- we will put a prime mark on K as a warning.)

44

HA

45

This equation states that _____ and _____ will be formed until their concentrations are large

Page 13

enough to cause the reverse reaction to go as fast as the _____ one.

46

The more _____ and _____ required to satisfy this equation, the stronger is our weak acid. Notice that a weak acid is one that in water solution dissociates only partially. We see that HCl, no matter how dilute, cannot be considered a _____, because it dissociates completely in water solution.

47

Solve the mass–law equation, $\dfrac{[H^+] \times [A^-]}{[HA]} = K'$, for $[H^+]$.

48

Restating the resulting equation in logarithmic language, we have:

$$\log [H^+] = \text{_____} + \log \frac{[HA]}{[A^-]}$$

49

Multiplying each term of this equation by minus one, we can obtain:

$$- \log [H^+] = \text{_____} + \log \frac{[A^-]}{[HA]}$$

Notice that the ratio $[HA]/[A^-]$ has been inverted to keep this term positive.

45

H^+, A^- (either order)

forward

46

H^+, A^- (either order)

weak acid

47

$$[H^+] = K' \frac{[HA]}{[A^-]}$$

48

$\log K'$

49

$-\log K'$

50

But $-\log [H^+] = pH$. By analogy we can let a new symbol, pK', stand for $-\log K'$. If we do this we can write

$$\underline{\hspace{2cm}} = \underline{\hspace{2cm}} + \log \frac{[A^-]}{[HA]}$$

50

(The full equation)

$$pH = pK' + \log \frac{[A^-]}{[HA]}$$

51

This is the *Henderson-Hasselbalch* equation: $pH = pK' + \log \dfrac{[A^-]}{[HA]}$. This equation $\boxed{\text{does/does not}}$ involve any new assumptions. It is $\boxed{\text{more/no more}}$ than a logarithmic form of the mass–law equation.

51

does not

no more

52

If K' is a constant, so is pK'. Therefore, the $\underline{\hspace{3cm}}$ equation, $pH = pK' +$ $\underline{\hspace{2cm}}$, shows how the pH of a solution will change as we increase the concentration of $\underline{\hspace{1.5cm}}$ and decrease the concentration of HA by adding NaOH.

52

Henderson–Hasselbalch

$$\log \frac{[A^-]}{[HA]}$$

A^-

53

For example, if we add enough NaOH to HA to convert 90% of it to A^- we shall have these two present in the ratio $\dfrac{[A^-]}{[HA]} = \dfrac{90}{10}$, or 9. The logarithm table shows that the logarithm of 9 is $\underline{\hspace{1cm}}$.

54

Therefore if this weak acid has a pK' of 5.00, the resulting pH will be 5.00 + _____ = _____ .

55

Using a log table, proceed in the same way to fill in the blank spaces in this table, which shows how $[A^-]/[HA]$ changes during a titration. Assume that this weak acid has a pK' = 5.00, and then calculate and tabulate in the last column the pH values predicted at each degree of neutralization.

% Neutralization	$\dfrac{[A^-]}{[HA]}$	$\log \dfrac{[A^-]}{[HA]}$	pH
1	1/99	-2.00	3.00
10	10/90	_____	_____
20	20/80	_____	_____
30	30/70	_____	_____
50	_____	_____	_____
70	70/30	_____	_____
80	80/20	_____	_____
90	90/10	0.95	5.95
99	99/1	_____	_____
99.9	999/1	_____	_____

53

0.95

54

0.95

5.95

55

$\dfrac{[A^-]}{[HA]}$	$\log \dfrac{[A^-]}{[HA]}$	pH
	-0.95	4.05
	-0.60	4.40
	-0.37	4.63
1.00	0.00	5.00
	+0.37	5.37
	+0.60	5.60
	+0.95	5.95
	+2.00	7.00
	+3.00	8.00

56

Plot these predicted pH values on the graph below and draw a smooth line through the points.

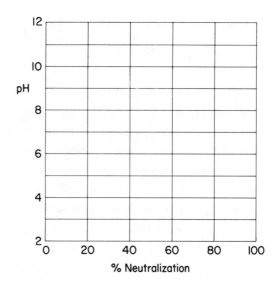

56

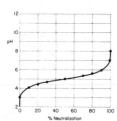

57

Notice that this theoretical curve, which we have computed from the mass–action law, has ⌐the same form as/a different form from⌐ that actually obtained in the laboratory, as pictured earlier in this program (see Item 32). Take note that our graphic representation of the relation between the pH and the % neutralization deserves to be drawn with some care. Of the three plots shown here, which corresponds to the Henderson-Hasselbalch equation? _____

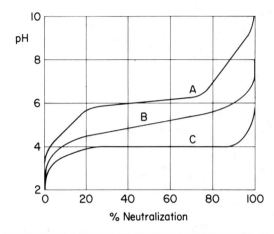

% Neutralization

58

In the table you completed in Item 55, we introduced a new assumption, namely that the amount of A^- present is given by the quantity of _____ added. This assumption is not strictly correct at the lower end of the titration curves, especially for stronger weak acids, since some _____ (and H^+) is formed by spontaneous dissociation.

59

By the same token, you cannot properly use the Henderson-Hasselbalch equation to calculate the pH at which an acid is 100% dissociated, by substituting as follows:

$$pH = pK' + \log \frac{100}{0}$$

57

the same form as

B (For a well-drawn curve, the portion falling between 9 and 91% neutralization approximates two pH units. A covers over twice that, C much less than that.)

58

OH^- (or NaOH)

A^-

As you will discover if you try to solve this equation, a pH at which [_____] is actually zero can never be reached.

59

HA

INDICATORS AND ENDPOINTS

60

To decide at what pH to terminate a titration in ordinary analytical work, we need to know only at what pH the weak acid will have been sufficiently titrated to yield the desired accuracy. For example, if we want accuracy within 0.1% (which is usually quite sufficient), we need to know at which pH HA will have been converted to A^- to the extent of 99.9%. We can see from the table in Item 55 that this occurs at a pH lying _____ above the pK'.

60

3 units

61

Hence, to secure an analytically complete titration we should terminate it when the pH has been brought about _____ units beyond the pK' of the weak acid being titrated. If we use an indicator for the purpose of recognizing the endpoint of a titration, the indicator should be one that will show a color change ⎹ at the pK'/about three units beyond the pK' ⎸ of the weak acid. Stopping two units above the pK will give us 99% accuracy.

62

But indicators do not show a sudden change as a certain pH is reached. Instead, they are also weak acids; but they are weak acids that have a conjugate base of a different color. If the weak acid form of the indicator is yellow and the weak base is red, as the pH is raised the color will progress from _____, through _____ hues, to _____ .

63

By mental use of the Henderson-Hasselbalch equation we can calculate that a solution of the indicator, phenol red, which has a pK$'$ of 7.9, will contain at pH 6.9 _____ times as many molecules of the yellow, undissociated form as it does of the red, _____ form. Therefore, the hue will be a | yellow-orange/orange-red | .

64

At pH 5.9 the fraction in the | yellow/red | form would be too small for observation. Would phenol red serve to differentiate between two solutions, one with a pH of 4 and one with a pH of 5? _____ Would it serve to differentiate between two solutions, one with a pH of 11 and one with a pH of 12? _____

61

three

about three units
beyond the pK$'$

62

yellow

orange

red

63

10

dissociated
(conjugate base)

yellow-orange

64

red

No

No

65

pK′

66

pK′

lies about three units
beyond

67

about three units be-
yond

65

We may conclude that indicators are useful in differentiating pH values only within the range of about one unit above and one unit below the _____ of the indicator.

66

Acid–base indicators are generally used for either of two purposes; to measure the pH, or to indicate when a titration is completed. Keeping distinct the different grounds for their selection for these two purposes is important. We have just seen that for the first application we select an indicator whose _____ is near the pH to be measured. On the other hand, when we are titrating a weak acid to see how concentrated it is, the pH we want to be able to recognize | equals/lies about three units beyond | the pK′ *of the weak acid.*

67

Putting together these two criteria, the pK′ of an indicator for titrating a weak acid in order to determine its strength should lie | at/about three units beyond | the pK′ of the weak acid.

68

If you were titrating acetic acid, pK′= 4.7, in order to learn its concentration, which indicator would you prefer: bromcresol green, pK′ = 4.7, or phenol red, pK′ = 7.9? _____ If, instead, you wanted to determine the pH of a solution, known to lie between pH 4 and 5, which indicator would you prefer? _____

69

In an ordinary titration one titrates the weak acid in question as a matter of course, but in order to obtain a visible color change at the endpoint, he must partly titrate the _____ also. If no significant error is to result from this fact, the indicator must be present in very _____ concentration; accordingly it must have the property of being | weakly/intensely | colored. Implicit in this use of indicators is the axiom that each conjugate pair in a given solution will show a ratio of their concentrations | identical with other conjugate pairs/consistent with the pH | .

68

phenol red

bromcresol green

69

indicator

low

intensely

consistent with the pH

BUFFERING

70

Because the pK' marks the flattest part of the titration curve, it is the point at which the pH changes the least on adding reasonable amounts of H^+ or OH^-. Added H^+ is largely consumed by reaction with the conjugate _____ , A^-. Added OH^- is consumed by reaction with the _____ , HA.

70

base

weak (conjugate)
acid

71

This action to minimize the change in pH is called *buffering*. Good buffering action will be expected only in the pH region around the _____ . By examining a titration curve, as reproduced below, we see that good _____ | would /would not | be obtained 0.5 pH units above or below the pK'.

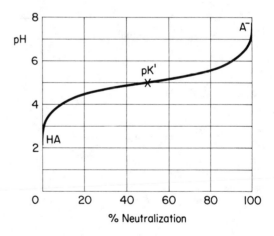

72

To minimize pH changes, that is, to secure effective _____ , we should therefore
select a weak acid whose _____ is near the
desired pH, and usually less than one unit removed
from that pH. This weak acid is combined with its
conjugate _____ , to form what we call a
buffer system.

73

The Henderson-Hasselbalch equation,

$$pH = pK' + \log \frac{[A^-]}{[HA]}$$

shows us how the pH of a given buffer solution is
determined by the opposed actions of the weak
acid in releasing _____ , and of the _____
_____ in binding _____ .

71

pK′

buffering

would

72

buffering

pK′

base

73

H$^+$

conjugate (weak) base

H$^+$

THE BICARBONATE-CARBONIC ACID SYSTEM

74

Consider, for example, that an important buffer stabilizing our plasma is the bicarbonate-carbonic acid system. Carbonic acid, H_2CO_3, like other dibasic and polybasic weak acids, dissociates only one H^+ at a time. The dissociation reaction yielding bicarbonate ion therefore is:

$$H_2CO_3 \rightleftharpoons \underline{\hspace{1cm}} + \underline{\hspace{1cm}} .$$

74

$H^+ + HCO_3^-$

(either order)

75

$$pH = pK' + \log \frac{[HCO_3^-]}{[H_2CO_3]}$$

As we can see from the Henderson–Hasselbalch equation, a particular value for the ratio, $[HCO_3^-]/[H_2CO_3]$, corresponds to a particular plasma pH. If we double the concentration of our conjugate acid, _____ , its H^+ donating action is thrown out of balance with the H^+ binding action of _____ the conjugate base.

75

H_2CO_3

HCO_3^-

76

Therefore, a small amount of the extra H_2CO_3 dissociates to lower the pH. Notice that relatively little free H^+ is present, so that proportionately very little of the added H_2CO_3 needs to dissociate to double the $[H^+]$. Thus we may neglect the small loss of _H_2CO_3_ and the small gain of HCO_3^- in this adjustment.

77

Or if, instead, the $[HCO_3^-]$ of the plasma is doubled, the pH will be ⌐raised/lowered⌐ because a little of this added *base* will take up H^+. From the mass–law equation

$$[H^+] = K' \cdot \frac{[H_2CO_3]}{[HCO_3^-]}$$

we can see that doubling the $[HCO_3^-]$ will cause the $[H^+]$ to be ⌐doubled/halved⌐ , assuming that the $[H_2CO_3]$ is at the same time essentially constant.

PHOSPHORIC ACID

78

Next, let us picture the titration of $0.1\,M$ H_3PO_4 with N NaOH. First we discover a region of dissociation, centered about pH 2.0, where the following reaction occurs:

$$H_3PO_4 \rightleftharpoons H^+ + \underline{\hspace{3cm}} .$$

Notice that only one H^+ is released at this stage so that only _____ equivalent(s) of NaOH per mole of H_3PO_4 will be needed to carry out this titration step.

76

H_2CO_3

77

raised

halved

78

$H_2 PO_4^-$

one

79

Draw on this graph a sigmoid curve to approximate the pH change expected during the addition of this *first* equivalent of NaOH. (Examine the pH range covered in the preceding titration curves [see Item 56] to approximate the correct shape.)

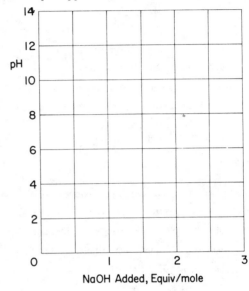

79

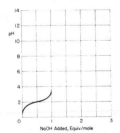

80

As the graph in Item 79 showed, the resistance to pH change falls to almost zero around pH 4.5; but then as we add more NaOH we encounter a second region of dissociation, centered around pH 6.8. The reaction is

$$H_2 PO_4^- \rightleftharpoons \rule{3cm}{0.4pt} + \rule{3cm}{0.4pt} .$$

Notice again that only one of the two remaining hydrogens is involved.

81

If this second pK', which we will call pK'_2, has value of 6.8* then in a solution at pH 6.8 we should have present _____ % $H_2PO_4^-$ and _____ % HPO_4^{2-}.

82

Notice that the mass–action law | does/does not | call for H_3PO_4 and PO_4^{3-} to be totally absent at pH 6.8, even though $H_2PO_4^-$ and _____ are the principal ionic species present and serve best to calculate the pH using pK'_2.

83

Indeed we can use the Henderson–Hasselbalch equation also to calculate the concentration of H_3PO_4 or PO_4^{3-}. We can write the equation for calculating the relationship between $[H_3PO_4]$ and $[H_2PO_4^-]$ at pH 6.8 as follows:

$$pH = pK' + \log \frac{[H_2PO_4^-]}{[H_3PO_4]}$$

To select the proper pK' to use in this case, we must consider which substance serves as the conjugate acid, and which serves as the conjugate base, in the fraction at the right of the equation. This conjugate pair then identifies the right pK'. In the present case this pair refers to the | first/ second/third | pK' of phosphoric acid.

*Don't be concerned if the value given by a textbook or other source is slightly different. This value is very sensitive to salt concentration, and will vary when conditions are slightly different.

80

$HPO_4^{2-} + H^+$

(either order)

81

50, 50

82

does not

HPO_4^{2-}

83

first

84

6.8, 2.0

0.001%

85

cannot

Henderson–Hasselbalch

84

Accordingly we can fill in the appropriate values for the pH and pK$'$ in order to calculate the H_3PO_4 concentration at pH 6.8, as follows:

$$\underline{\hspace{3cm}} = \underline{\hspace{3cm}} + \log\frac{[H_2PO_4^-]}{[H_3PO_4]}$$

On solving this equation, we can see that at pH 6.8 the concentration of H_3PO_4 will be roughly $\boxed{1\%/0.1\%/0.001\%}$ of the concentration of $H_2PO_4^-$. We will use the Henderson–Hasselbalch equation to calculate $[PO_4^{3-}]$ shortly.

85

Suppose next that we add 18 ml of 0.1 N NaOH to 10 ml of 0.1 M H_3PO_4. How do we determine which pK$'$ to use? Certainly we $\boxed{\text{can/cannot}}$ properly insert the numbers 18 and 10 as numerator and denominator directly into the _____ equation. First we must consider the effect of a chemical reaction produced by the added NaOH. Always examine first for this possibility!

86

Since the unit of normality can be stated as milliequivalents/ml, the 18 ml will contain 18 ml $\times$ 0.1 mE/ml = _____ of OH^-.

87

Of these, exactly _____ mE of OH^- will be consumed in converting the 1.0 millimole of H_3PO_4 into $H_2PO_4^-$. The remaining 0.8 mE of OH^- will then react with some of the $H_2PO_4^-$ produced to form _____ mmole of HPO_4^{2-} and leave 0.2 mmole of _____ .

88

Since we are dealing now with the relationship between the concentrations of HPO_4^{2-} and _____ , we must use pK_2'. Hence, inserting the calculated values for these concentrations, we can write:

$$pH = 6.8 + \log \underline{\hspace{2cm}}$$

$$pH = \underline{\hspace{3cm}}$$

89

Bearing in mind that plasma normally has a pH of 7.4, what is the ratio of its $[HPO_4^{2-}]$ to the $[H_2PO_4^-]$? _____

90

Draw on the graph below a second sigmoid curve to approximate the pH change predicted during the second dissociation. Add a third sigmoid section to represent a pK_3' of about 11.8.

86

1.8 milliequivalents

87

1.0

0.8

$H_2PO_4^-$

88

$H_2PO_4^-$

0.8/0.2 = 4

pH = 6.8 + 0.6 = 7.4

89

Since

$$\log \frac{[HPO_4^{2-}]}{[H_2PO_4^-]} = 0.6,$$

$$\frac{[HPO_4^{2-}]}{[H_2PO_4^-]} = 4$$

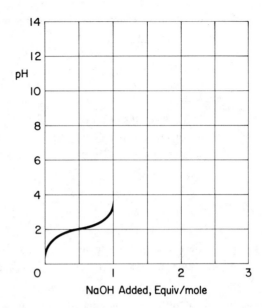

NaOH Added, Equiv/mole

90

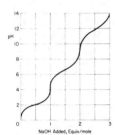

NaOH Added, Equiv/mole

91

Write the dissociation reaction that pK_3' represents: _____

91

$$HPO_4^{2-} \rightleftarrows PO_4^{3-} + H^+$$

92

Considering this titration curve, which of the indicators listed in the box below would you select as most nearly suitable in determining the strength of H_3PO_4 by titration with alkali: (a) for a titration using only one equivalent of NaOH per mole of H_3PO_4? _____ (b) for a titration using two equivalents of NaOH per mole of H_3PO_4? _____

Bromcresol green, pK' = 4.7
Chlorophenol red, pK' = 6.0
Phenol red, pK' = 7.9
Phenolphthalein, pK' = 9.7

(The selected indicators do not quite meet the criterion that their pK's should lie 3 units beyond the _____ of the weak acid group to be titrated. In case of difficulty, see again Items 60 to 68.)

93

Set up the Henderson–Hasselbalch equation relating the $[PO_4^{3-}]$, the $[HPO_4^{2-}]$, and the pH. For H_3PO_4, $pK_1' = 2.0$; $pK_2' = 6.8$; $pK_3' = 11.8$.

94

The concentration of HPO_4^{2-} in human blood plasma is about 0.001 M, and the pH, 7.4. Substitute these values into the equation you just wrote, and calculate the plasma concentration of tertiary phosphate ion, PO_4^{3-}.

92

Bromcresol green

Phenolphthalein

pK'

93

$$pH = 11.8 + \log \frac{[PO_4^{3-}]}{[HPO_4^{2-}]}$$

94

$4 \times 10^{-8} M$

Method:

$$7.4 = 11.8 + \log\frac{[PO_4^{3-}]}{0.001}$$

$$\log\frac{[PO_4^{3-}]}{0.001} = -4.4$$

$$[PO_4^{3-}] = 0.00000004 M$$

$$= 4 \times 10^{-8} M$$

95

third

96

cannot

95

This calculation shows that great difficulty attends the withdrawal of the | second/third | H^+ from H_3PO_4 and that it is dissociated only to a tiny extent at physiological pH values. Nevertheless, this H^+ must be removed before the characteristic $Ca_3(PO_4)_2$ structure of bone and tooth can be formed.

96

From examining the titration curve of phosphoric acid we | can/cannot | tell whether the solution titrated contains three different weak acids, or a single acid with three dissociable groups.

APPLICATION OF THE HENDERSON-HASSELBALCH EQUATION

97

The Henderson–Hasselbalch equation is often used to calculate the pH of buffer systems. In doing this, we must distinguish carefully between two different practical situations. First, we can prepare a buffer mixture by mixing 10 ml of a 0.1 N weak acid, pK' 5.0, with 5 ml of a 0.1 N solution of a salt of the _____ , Na^+A^-. (The solution of the salt of the weak acid of course contains 0.5 milliequivalents of A^-.) This combination yields a ratio $[A^-]/[HA] =$ _____ , and therefore the pH $= 5.0 + ($ _____ $) =$ _____ .

98

In contrast, we may prepare a buffer by adding instead 5 ml of 0.1 N NaOH to the 10 ml of the weak acid. Notice that in this case all of our A^- is generated from HA by reaction with NaOH. Now the fraction $[A^-]/[HA]$ will have the value _____ , and the pH will be _____ . (If this exercise is difficult for you, repeat the calculations of Item 55. A similar calculation would apply if you were to prepare a buffer from sodium acetate and HCl.)

THE pK' AND THE IONIC STRENGTH

99

Up to this point we have apparently assumed that the pH of a buffer depends entirely on the ratio of $[A^-]$ to $[HA]$. In other words, we have treated the _____ as a constant. If this were precisely correct, a one-molar buffer adjusted to a given pH

97

(weak) acid

1/2

−0.3, 4.7

98

5/5 or 1

5.0

could be diluted to any extent we liked without changing the _____ .

99

pK′ (or pK, K′)

pH (or [H⁺])

100

ionic strength (ionic
concentration; salt
concentration)

100

In actuality we know that a buffer can be diluted only moderately without excessive change in its pH. The changes in pH with total salt concentration are especially rapid at salt concentrations above $0.1M$. We know that ions are restrained in their free movement at high total ion concentrations* (expressed best as the *ionic strength*). Furthermore, a *divalent* ion, such as HPO_4^{2-}, is restrained more strongly than a *univalent* one, like $H_2PO_4^-$. Therefore a 50:50 mixture of these two ions will yield a substantially lower pH at a high

_____ .

101

Hence we see that the value of pK′ changes as the ionic strength is changed. If we trace the course taken by the value of pK′ at lower and lower ionic strengths, we may ultimately discern the limiting value applying for an *infinitely dilute* solution. For this value of pK′ we may omit the prime mark and write simply _____ . The same value should also apply at any ionic strength if we obtain and use the *activities* of HA and A⁻, rather than

*For example, the conductivity of a 0.1 N NaCl solution is somewhat less than ten times that of a 0.01 N solution. The apparent concentration of Na⁺ or Cl⁻ is somewhat less than the actual concentration, because of this effect of other ions on the movement of each ion in more concentrated solutions.

Page 35

their molar concentrations. Therefore we may write:

$$pH = \underline{\hspace{1cm}} + \log \frac{a_{A^-}}{a_{HA}} , \text{ and}$$

$$pH = \underline{\hspace{1cm}} + \log \frac{[A^-]}{[HA]}$$

The ⎡first/second⎤ equation applies at any ionic strength; the other equation is rigorously correct only for a specified ionic strength.

102

You will sometimes see tables that record two pK values for a given system, e.g., for phosphoric acid, pK_2 = 7.2; pK_2' = 6.8. The first symbol, \underline{\hspace{1cm}}, represents the value applying to the second dissociation of phosphoric acid in an infinitely \underline{\hspace{1cm}} solution; \underline{\hspace{1cm}} is the symbol for the value applying to the same dissociation at the ionic strength of the blood plasma, or in other cases, for some other arbitrary ionic strength.

103

This device of using a different value for the pK′ for each set of conditions might suggest that only the degree of dissociation changes, and nothing else. Actually the device is only a convenient way to avoid correcting separately the concentration of each participating substance by instead attributing all changes to change in the \underline{\hspace{2cm}} constant.

101

pK

pK

pK′

first

102

pK_2

dilute

pK_2'

103

dissociation

(ionization)

THE AMMONIUM ION

104

To explore the nature of weak acids further, let us picture the titration of 50 ml of 0.1 N NH_4Cl. If we add HCl we detect no structure that will bind _____. But by adding NaOH we find present a substance that can dissociate H^+. This substance cannot be Cl^-; it must be _____ .

104

H^+

NH_4^+

105

The dissociation reaction is:

$$NH_4^+ \rightleftharpoons \text{_____} + \text{_____} .$$

In this case _____ is the weak acid, and _____ is its _____ base.

105

$NH_3 + H^+$
(either order)

NH_4^+

NH_3

conjugate

106

The pK′ of this dissociation is about 9.4. Trace at the appropriate position on the graph below the approximate form of the titration curve you would expect for titration of the 50 ml portion of 0.1 M NH_4Cl.

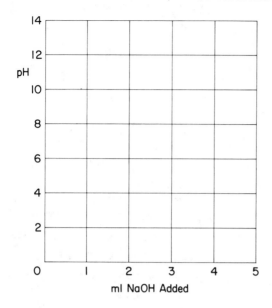

ml NaOH Added

107

The statement that the pK′ is 9.4 means experimentally that the dissociation centers symmetrically around pH _____ . Accordingly, NH_4^+ dissociates its H^+ only when the $[H^+]$ falls to about $10^{-9}\,M$. To permit NH_4^+ to dissociate, the $[H^+]$ must be brought much lower than would be necessary to permit acetic acid to dissociate. Therefore, ammonium ion is the ⌐weaker /stronger⌐ acid than acetic acid.

108

Conversely, NH_3 has a ⌐greater/smaller⌐ tendency to *associate* H^+ than has acetate. Therefore, NH_3 is the ⌐stronger/weaker⌐ base.

106

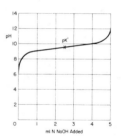

107

9.4

weaker

108

greater

stronger

109

base

conjugate acid

$$pH = pK' + \log \frac{[NH_3]}{[NH_4^+]}$$

110

acid

alkaline (basic)

H^+

109

We can now write the Henderson–Hasselbalch equation for the NH_4^+ – NH_3 system, being careful to put the concentration of the conjugate _____ in the numerator and that of the _____ in the denominator, as follows: _____ .

110

Ammonium chloride solution is acidic because it contains a weak _____ ; similarly, sodium acetate solution is _____ because it contains a weak base, the acetate ion, which binds _____ .

UNITARIAN CONSIDERATION OF ACIDS AND BASES

111

You may have been taught earlier to show NH_3 hydrated in solution to form NH_4OH. You should come to understand, however, that NH_3 and acetate ion are really weak _____ . even though they contain no hydroxide groups.

112

The great advantage of representing these dissociations as

$$RNH_3^+ \rightleftharpoons H^+ + \underline{\hspace{1cm}}$$

is that we always visualize the acid releasing or donating _____ , and the conjugate base as accepting or binding _____ .

113

What we are doing here is always to consider $\boxed{H^+/OH^-}$ as a direct reactant, whether we are discussing the behavior of a weak acid or that of a weak base. If we insist on showing weak bases as dissociating OH^- rather than binding _____ , we make the whole story much more complicated and no longer get easily compared, parallel titration curves.

114

Below is a summary list of the chemical equations we have used to illustrate H^+ dissociation. The substances on the left, namely the weak _____ , are listed in order of their $\boxed{\text{increasing/decreasing}}$ strength. The substances on the right are conjugate _____ listed in order of their _____ strength.

111

bases

112

RNH_2

H^+

H^+

113

H^+

H^+

$$H_3PO_4 \rightleftharpoons H^+ + H_2PO_4^- \qquad pK' = 2.0$$
$$HOAc \rightleftharpoons H^+ + OAc^- \qquad pK' = 4.7$$
$$H_2CO_3 \rightleftharpoons H^+ + HCO_3^- \qquad pK' = 6.1$$
$$H_2PO_4^- \rightleftharpoons H^+ + HPO_4^{2-} \qquad pK' = 6.8$$
$$NH_4^+ \rightleftharpoons H^+ + NH_3 \qquad pK' = 9.4$$
$$HCO_3^- \rightleftharpoons H^+ + CO_3^{2-} \qquad pK' = 9.8$$
$$HPO_4^{2-} \rightleftharpoons H^+ + PO_4^{3-} \qquad pK' = 11.8$$

114

acids

decreasing

bases

increasing

115

anions

anions (charged)

charge

116

HPO_4^{2-}

H_3PO_4

acid ⎫
        ⎬ (either order)
base ⎭

115

Some of the weak acids in this list are uncharged substances; others are cations; still others are _____ . Similarly, one of the weak bases listed is an uncharged substance, whereas the others are _____ . Discussion would only be complicated by dividing the list into two or three lists for separate treatment because of these differences in _____ .

116

Notice that $H_2PO_4^-$ appears in the list in Item 114 once as an acid and once as a base. It is the conjugate acid to _____ and the conjugate base to _____ . Unless we designate the conjugate partner to which it is being referred, we cannot say arbitrarily that $H_2PO_4^-$ is a(n) _____ , or that it is a(n) _____ .

AMMONIUM ACETATE AND GLYCINE

117

Next let us picture the titration of a solution of ammonium acetate. On adding NaOH we will again discover the dissociation of _____ from NH_4^+, which we encountered earlier for NH_4Cl. The pK' for this dissociation, as we have observed, is 9.4. If, instead of NaOH, we add HCl to the original ammonium acetate solution, we note the uptake of H^+ by _____ . This reaction, you will recall, has a _____ of 4.7.

118

The complete titration for 50 ml of the solution of 0.1 M ammonium acetate with N NaOH and N HCl can be charted by joining the curves traced earlier for each of these two titrations (Items 56 and 106). On the graph below sketch the approximate appearance of the overall titration curve, using the pK' values already given.

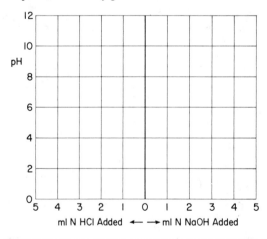

117

H^+

acetate

pK'

118

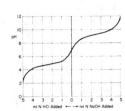

119

Mark on this curve:

An I to show the point representing the initial pH.

An A^- to represent the composition NH_3 + CH_3COO^-.

A B^+ to represent the composition NH_4^+ + CH_3COOH.

pK_1' to represent the point at which equal quantities of CH_3COOH and CH_3COO^- will be present.

pK_2' to represent the point at which equal quantities of NH_4^+ and NH_3 will be present.

119

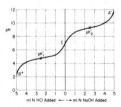

120

Recalling that the resistance to pH change is the greatest where the curve is the flattest, we can see that the buffering is $\boxed{\text{strong/weak}}$ near the point pK_2', but _____ at point I.

120

strong

weak

121

Use the relationship, $pH = 9.4 + \log \dfrac{[NH_3]}{[NH_4^+]}$, to calculate what fraction of the NH_4^+ will take the form, NH_3, in a solution at pH 7.05.

122

Similarly, use the relationship,

$$pH = 4.7 + \log \frac{[acetate]}{[acetic\ acid]}$$

to calculate how high the $[CH_3COOH]$ should be at this pH, in relationship to the $[CH_3COO^-]$. (There should be _____ as much acetic acid as acetate.)

123

Accordingly, these calculations show that some _____ and some _____ are present in ammonium acetate, but that over 99.5% of the substance is in the form $[NH_4^+]$ $[CH_3COO^-]$. If you have the opportunity, sniff some crystalline ammonium acetate. Note that you can smell simultaneously both _____ and _____ .

121

The answer is: $\dfrac{1}{224}$

Our calculations:

$$7.05 = 9.4 + \log \frac{[NH_3]}{[NH_4^+]}$$

$$\log \frac{[NH_3]}{[NH_4^+]} = -2.35$$

2.35 is the log of 224. Hence, $[NH_3]$ would be only $\dfrac{1}{224}$ of $[NH_4^+]$.

122

The answer is: $\dfrac{1}{224}$

Our calculations:

$$7.05 = 4.7 + \log \frac{[A^-]}{[HA]}$$

$$\log \frac{[A^-]}{[HA]} = 2.35$$

Hence $[CH_3COOH]$ would be only $\dfrac{1}{224}$ of $[CH_3COO^-]$.

123

ammonia $\Big\}$ (either

acetic acid $\Big/$ order)

ammonia $\Big\}$ (either

acetic acid $\Big/$ order)

124

would accept

(Do not be con-

cerned if your prefer

another form at this

point.)

124

Suppose that the NH_4^+ and CH_3COO^- were joined together by a covalent link, thus

$$H_3^+N - CH_2COO^-.$$

This substance is glycine, or aminoacetic acid. Would you accept this structure as written or prefer another? _____

125

The figure below represents the dissociation curve for glycine (solid line), drawn using $pK_1' = 2.3$ and $pK_2' = 9.7$, superimposed on the titration curve (dashed line) for ammonium acetate. (The distortion expected for a titration curve below pH 3 has been corrected.) Mark I for the initial point of the titration, G^- for the point where the composition is mainly $NH_2 - CH_2COO^-$, and G^+ where it is mainly $H_3^+N - CH_2COOH$. Indicate where on the titration curve pK_1' and pK_2' lie.

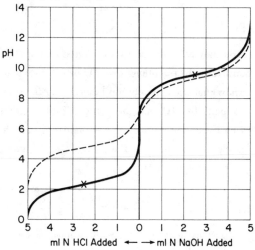

126

If a glycine solution were in an electric field, would the organic molecule tend to go to the anode or cathode at point G^- in the diagram? _____

127

At point I in the diagram, the tendency to go to the anode and cathode should be _____ .

128

Such a point is known as the *isoelectric point.* Now, we could write two structures for glycine that have no net charge and would therefore not migrate in an electric field (please complete):

$$NCH_2COO \qquad\qquad NCH_2COO$$

129

You may ask why have we preferred one of these structures to the other? From their pK' values we know that the amino group has a greater tendency to bind the H^+ than does the carboxyl group. Hence in the structure predominating at the isoelectric point, it is the _____ group that will retain the H^+.

130

We can calculate from the pK' values, as we did earlier for ammonium acetate, just how much of

---

125

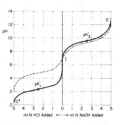

126

anode

127

equal (zero)

128

H_2NCH_2COOH

$H_3^+NCH_2COO^-$

(either order)

129

amino

the carboxyl group will be in the form $-COOH$ at point I, and how much of the amino group in the form $-NH_2$. Since both pK_1' and pK_2' are further removed from point I than was the case for ammonium acetate, even ⌐smaller/larger⌐ proportions of each will be present in the uncharged forms.

130

smaller

131

small

131

If we consider how many glycine molecules will have their carboxyl group in the form $-COOH$ and *at the same time* have their amino group in the form $-NH_2$, we can see that this number will be very ⌐small/large⌐ .

132

Of course we shall also have present at the isoelectric point a small amount of the anionic form of glycine:

NCH_2COO (Complete)

But as long as we also have present a similar amount in the cationic form,

NCH_2COO (Complete)

to balance the tendency of the anion to go to the anode, we shall encounter no net tendency to migration in an electric field, so that the requirements of the _____ point are still met.

133

Accordingly, write here the best single structure for glycine at the isoelectric point.

134

This structure is known as a *zwitterion*, or as a *dipolar* structure. Many molecules of biological importance are _____ or multi-polar.

GLUTAMIC ACID

135

Here is the generalized structural formula for an amino acid:

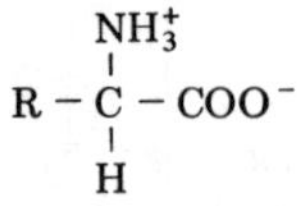

$$R - \overset{\overset{\displaystyle NH_3^+}{|}}{\underset{\underset{\displaystyle H}{|}}{C}} - COO^-$$

Where the sidechain R of the amino acid as illustrated here includes a third dissociating group, the titration curve must also show an additional _____ . One important amino acid for which R includes a dissociating group is glutamic acid, α-aminoglutaric acid. (Glutaric acid

132

$H_2NCH_2COO^-$

$H_3^+NCH_2COOH$

isoelectric

133

$H_3^+NCH_2COO^-$

134

dipolar
(zwitterions)

is $COOH \cdot CH_2 \cdot CH_2 \cdot CH_2 \cdot COOH$.) Glutamic acid shows $pK_1' = 2.2$; $pK_2' = 4.3$; _____ = 9.7. The latter obviously pertains to the _____ group.

135

dissociation
(step, stage)

pK_3'

amino

136

Suppose that we were to titrate glutamic acid, beginning with the structure,

$$
\begin{array}{c}
COOH \\
| \\
CH_2 \\
| \\
CH_2 \\
| \\
H-C-NH_3^+ \\
| \\
COO^-.
\end{array}
$$

Mark on the graph below the three pK' values, 2.2, 4.3, and 9.7; one for each step in the titration. Through each point trace a sigmoid curve of the usual slope. Join these curves into a single smooth curve.

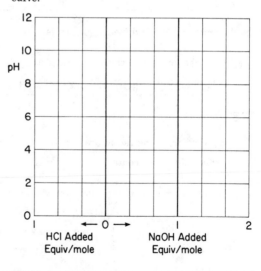

137

Mark I (isoelectric point), pK_1', pK_2' and pK_3' at appropriate points on the graph below. Indicate points at which the principal molecular species has a charge of -1, -2 and $+1$ by writing these numbers on the graph.

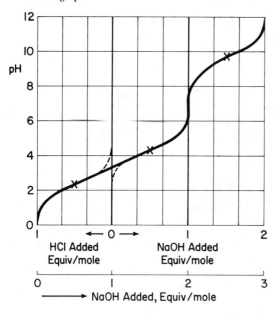

(Note that one could arrange to trace the whole curve using NaOH, as shown by the lower scale.)

138

Complete here the structure of the form of glutamic acid with a charge of -1.

$$
\begin{array}{c}
C \\
| \\
CH_2 \\
| \\
CH_2 \\
| \\
H - C - N \\
| \\
C
\end{array}
$$

136

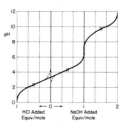

137

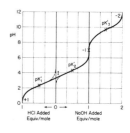

138

$$COO^-$$
$$|$$
$$CH_2$$
$$|$$
$$CH_2$$
$$|$$
$$H - C - NH_3^+$$
$$|$$
$$COO^-$$

139

$$COOH$$
$$|$$
$$CH_2$$
$$|$$
$$CH_2$$
$$|$$
$$H - C - NH_3^+$$
$$|$$
$$COOH$$

+1

140

$$COOH$$
$$|$$
$$CH_2$$
$$|$$
$$CH_2$$
$$|$$
$$H - C - NH_3^+$$
$$|$$
$$COO^-$$

3.25

Yes

139

Write here the structural formula for the form of glutamic acid tending most strongly to go to the cathode:

This substance is represented on the graph by the number _____ .

140

Complete here the formula for the form of glutamic acid predominating at the isoelectric point:

$$COOH$$
$$|$$
$$CH_2$$
$$|$$
$$CH_2$$
$$|$$
$$H - C - N$$
$$|$$
$$C$$

Since as a first approximation the isoelectric point will lie midway between pK_1' and pK_2', its value will be about _____ . Is this about the pH you would expect if you were to dissolve crystalline glutamic acid in water? _____

LYSINE, HISTIDINE, AND ARGININE

141

Another important amino acid having a third dissociating group is lysine, α, ϵ-diaminocaproic acid, caproic acid being the normal 6-carbon fatty acid. This amino acid shows $pK_1' = 2.2$, $pK_2' = 9.0$, $pK_3' = 10.5$. Of these, pK_1' must pertain to the _____ group.

142

Starting both the titration with NaOH and that with HCl with the structure shown,

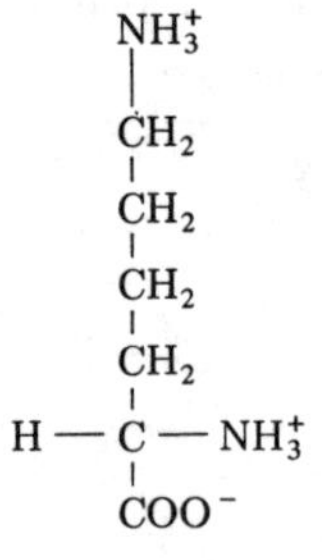

$$
\begin{array}{c}
NH_3^+ \\
| \\
CH_2 \\
| \\
CH_2 \\
| \\
CH_2 \\
| \\
CH_2 \\
| \\
H-C-NH_3^+ \\
| \\
COO^-
\end{array}
$$

mark on this graph the three pK' values. Draw lightly through these points sigmoid curves of the usual shape. Join these together into a single smooth curve.

141

carboxyl

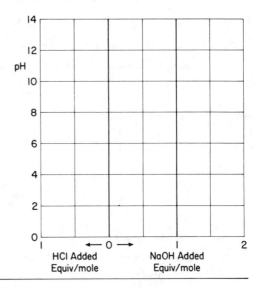

HCl Added
Equiv/mole

NaOH Added
Equiv/mole

142

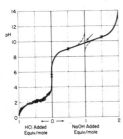

HCl Added
Equiv/mole

NaOH Added
Equiv/mole

143

Mark on the curve, I, pK_1', pK_2' and pK_3' at the appropriate points. Indicate by writing the numbers on the curve points at which the principal molecular species will have a net charge of -1, of $+1$, of $+2$.

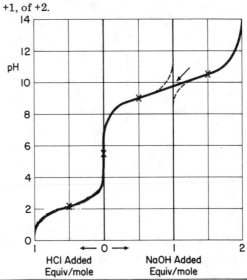

HCl Added
Equiv/mole

NaOH Added
Equiv/mole

144

Complete below the formula of the form of lysine predominating at the isoelectric point:

$$
\begin{array}{c}
CH_2\,N \\
| \\
CH_2 \\
| \\
CH_2 \\
| \\
CH_2 \\
| \\
H - C - NH_2 \\
| \\
C
\end{array}
$$

Since the isoelectric point will be expected to lie about midway between $pK_2' = 9.0$ and _____ = 10.5 it should have a value of about _____ .

145

Write here the structural formula for the form of lysine having a net charge of +2:

This species will predominate in a solution that is

$\boxed{\text{strongly acid/strongly alkaline}}$.

143

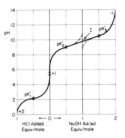

144

$$
\begin{array}{c}
CH_2\,NH_3^+ \\
| \\
CH_2 \\
| \\
CH_2 \\
| \\
CH_2 \\
| \\
H - C - NH_2 \\
| \\
COO^-
\end{array}
$$

pK_3'

9.75

145

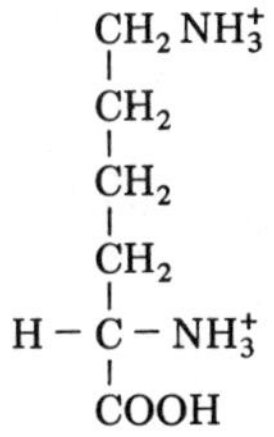

$$CH_2NH_3^+$$
$$|$$
$$CH_2$$
$$|$$
$$CH_2$$
$$|$$
$$CH_2$$
$$|$$
$$H-C-NH_3^+$$
$$|$$
$$COOH$$

strongly acid

146

$$CH_2NH_2$$
$$|$$
$$CH_2$$
$$|$$
$$CH_2$$
$$|$$
$$CH_2$$
$$|$$
$$H-C-NH_2$$
$$|$$
$$COO^-$$

strongly alkaline

146

Write here the structure of the form of lysine tending to go to the anode:

This species will predominate in a solution that is strongly acid/strongly alkaline .

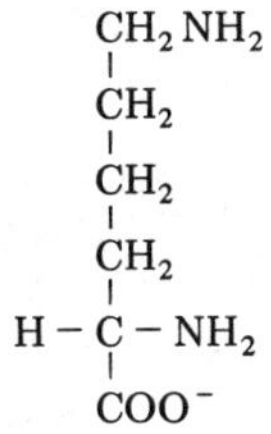

147

As another example, histidine has the following structure:

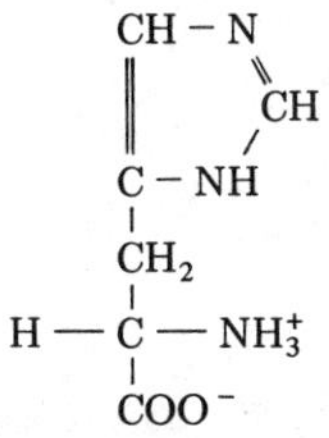

$$
\begin{array}{c}
CH-N \\
\| \quad\quad \diagdown CH \\
C-NH \diagup \\
| \\
CH_2 \\
| \\
H-C-NH_3^+ \\
| \\
COO^-
\end{array}
$$

The 5-membered heterocyclic ring, called the imidazole ring, will bind *one* hydrogen ion (it is immaterial which N is shown as already carrying H, and which is shown as receiving H^+) with a pK' of 6.0. The structure sketched is the one that will predominate above/below pH 6.

148

Using the procedure we have employed before, draw an approximate dissociation curve for histidine on this graph, using $pK_1' = 1.8$; $pK_2' = 6.0$; and $pK_3' = 9.2$. Mark the curve as usual.

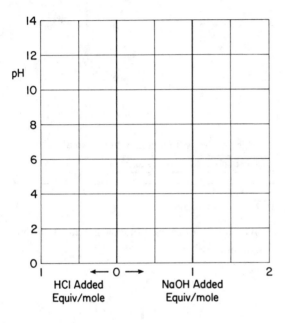

147

above

149

The abscissa scale shown here indicates that we began our titration by making a solution of [histidine monohydrochloride/histidine dihydrochloride] . Suppose that instead the titration was begun by making a solution of crystalline histidine, and adding NaOH or HCl as needed to trace the curve. Draw below the present abscissa scale another parallel scale to represent the equivalents of each reagent added per mole. We are reminded

148

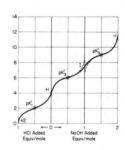

that titration curves can be traced experimentally beginning at any selected point.

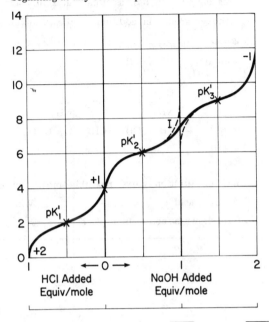

149

histidine
monohydrochloride

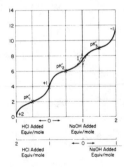

150

Other amino acids carry special dissociating groups on their sidechains. Arginine carries the guanidinium group.

$$\overset{\overset{+}{NH_2}}{\underset{\|}{NH_2 - C - NH -}}$$

which releases a H$^+$, showing a pK' of 13. (Despite the presence of 3 nitrogen atoms in the guanidinium group, only one of them dissociates a H$^+$.) The form shown here predominates $\boxed{\text{above/below}}$ pH 13.

151

This amino acid has, as usual, an amino and a carboxyl group, in addition to the guanidinium group:

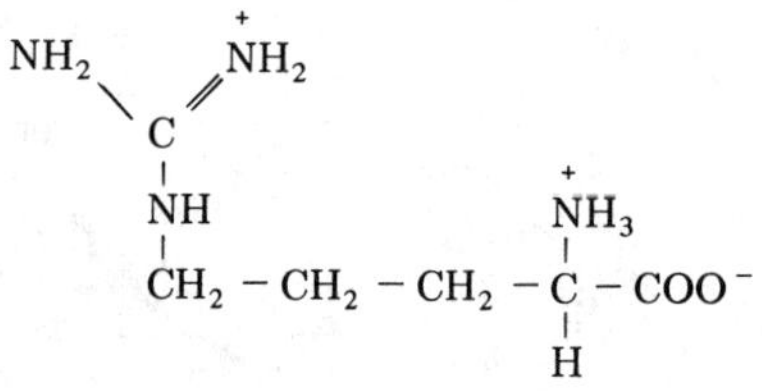

$$NH_2 \diagdown \overset{+}{C} \diagup \overset{+}{NH_2}$$
$$|$$
$$NH$$
$$|$$
$$CH_2 - CH_2 - CH_2 - \overset{\overset{+}{NH_3}}{\underset{H}{C}} - COO^-$$

Will its titration curve resemble most that of glutamic acid or lysine? _____ Will the form shown be abundant at pH 7? _____

150

below

151

lysine

yes

PEPTIDES

152

We turn now from the titration of the amino acids to the titration behavior of peptides and proteins, compounds formed by linking amino acids together. (A preliminary study of their structure must be presumed at this point. If you are totally unfamiliar with their structure, stop to study a preliminary textbook account.)

Peptides are formed by the joining of *amino acids* into chains by *peptide links*. In this structure, alanylglycine, underscore heavily the peptide link.

$$\overset{+}{N}H_3\,CH - C \overset{\displaystyle O}{\overbrace{}} NH - CH_2 - COO^-$$
$$\underset{CH_3}{|}$$

152

$$\overset{+}{H_3}\,NCHC - NHCH_2\,COO^-$$
$$\underset{CH_3}{|}$$

153

Here we show the structure for the tripeptide glycylglycylglycine.

$$H_3^+NCH_2\,\overset{\displaystyle O}{\overset{\|}{C}} - NHCH_2\,\overset{\displaystyle O}{\overset{\|}{C}} - NHCH_2\,COO^-$$

Would it be correct to say that glycylglycylglycine contains three glycine *molecules*? _____ Instead, we say that the peptide molecule contains three glycine *residues*. For example $(-NH - CH_2\,CO -)$ is a glycine _____ .

153

No

residue

154

Such peptides have titration curves somewhat like those shown by the free _____, except that the pK$'$ of the terminal carboxyl group is somewhat higher, the pK$'$ of the terminal amino group somewhat lower. The carboxyl and _____ groups involved in the peptide link no longer have a free existence. The atoms forming the peptide link do not readily dissociate or combine with a hydrogen ion.

154

amino acids

amino

155

Imagine a peptide composed of 100 alanine residues in a chain. Since the molecular weight of alanine is 89, and since elements of water have been lost in forming each peptide link, this peptide will have a molecular weight of 8900 – (99 X 18) = 7118. What weight of this peptide would bind (or release) one equivalent of hydrogen ion? _____ For comparison, what is the gram equivalent weight of alanine? _____

155

7118 g

89 g

PROTEINS

156

The peptide chains in proteins average certainly more than 100 amino acid _____ in length. This example suggests that on such chains, the free terminal _____ group and the free terminal _____ group contribute only minimal H^+ donating and accepting capacity.

157

Suppose we have such a peptide of 100 units, composed not entirely of alanine residues, but instead of alanine, leucine, lysine and glutamic acid residues in repeating sequence. Obviously in such a peptide the carboxyl and amino groups on the sidechains of glutamyl and lysyl residues will contribute much | more/less | to the titration

156

residues

amino ⎫
⎬(either order)
carboxyl ⎭

behavior of the peptide than will the terminal carboxyl and _____ group.

157

more

amino

158

carboxyl

lysine residues

158

This conclusion applies to proteins in general, because a large fraction, often as much as half, of the sidechains (represented by R in our general amino acid structure) contain dissociating groups; for example, _____ groups contributed by glutamic acid residues, or amino groups contributed by _____ . This feature is important to almost all the properties of proteins.

159

Here we consider the behavior of a typical protein. The table below shows the dissociating groups found in the protein, lactoglobulin, by both chemical analysis and examination of its titration curve. The amino acid residue showing the largest discrepancy between the two methods is the _____ residue.

Groups	Amino acid residues responsible	By analysis	By titration
carboxyl (β,γ)	glutamic and aspartic	60	60–63
carboxyl (α)	terminal	3	
amino (ϵ)	lysine	33	35–37
amino (α)	terminal	3	
imidazolium	histidine	4	6
guanidinium	arginine	7	5–7
TOTAL		110	108–111

160

The titration curve of such a protein is extremely complex, representing a total of about _____ dissociating groups. Nevertheless, the contribution of the several different classes of dissociating groups to the titration curve could be distinguished.

159

histidine

160

110

DISSOCIATION EXTENDED TO BINDING OF SMALL MOLECULES BY PROTEINS

(The next five items could well be studied initially, or reiterated, with Part II, *Enzyme Kinetics.*)

161

Dissociation curves serve equally well for describing the dissociation of ions other than H^+ and of small molecules from their complexes, for example with proteins, or conversely the association reactions by which these complexes are formed. Suppose the accompanying curve pertains to the following dissociation reaction:

$$Ca\ albuminate \rightarrow Ca^{2+} + albuminate^{2-}.$$

The abscissa shows the fraction of completeness of the formation of the complex. Label the ordinate scale with an expression for the calcium ion concentration in a form parallel to the pH as an expression of hydrogen ion concentration.

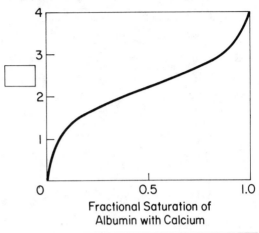

Fractional Saturation of
Albumin with Calcium

161

The ordinate should be labeled pCa^{2+}, or less ideally, $-\log [Ca^{2+}]$.

162

This extension of the logarithmic plot probably does not prove especially ingratiating to you; furthermore we can avoid it for our present purposes in favor of a non-logarithmic plot.

The curve drawn below shows the fractional dissociation of a weak acid, HA with $pK' = 5$, as a function of $[H^+]$ rather than of the pH. We see that the curve for such simple dissociations when plotted on a non-logarithmic scale has the form of a rectangular hyperbola.

The same curve has been arranged also to represent the binding of O_2 to myoglobin

$$Mb + O_2 \rightleftharpoons MbO_2$$

as a function of the oxygen concentration, which is plotted at the top of the graph in the form $\boxed{mM\ O_2 / \text{pressure of oxygen}}$.

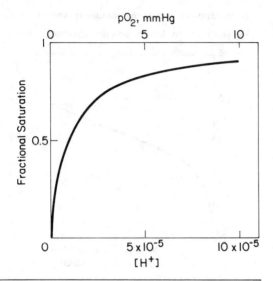

163

All the sigmoid titration curves we have plotted so far in Part I would appear as rectangular hyperbolas on a plot of this type. This style of plotting has the profound disadvantage, however, of requiring us to show hydrogen ion concentrations from perhaps 10^{-2} to $10^{-12}\,N$ for comparison of the various H^+ dissociation reactions that have interested us here. If one millimeter were selected to represent the interval from 10^{-11} to $10^{-12}\,N$, then the graph would need to be over 100 kilometers wide to show the whole range 10^{-2} to $10^{-12}\,N$! But when we are studying association or dissociation reactions whose equilibrium constants do not range widely, this plot is undoubtedly simpler to understand.

Here we compare graphically the reaction of oxygen and myoglobin with the reaction of the same gas and hemoglobin in an aqueous solution

162

pressure of
oxygen

resembling the inside of the human red blood cell. We see that the latter curve $\boxed{\text{is also a rectangular}}$ hyperbola, although corresponding to a different equilibrium constant/$\boxed{\text{has instead a sigmoid form}}$

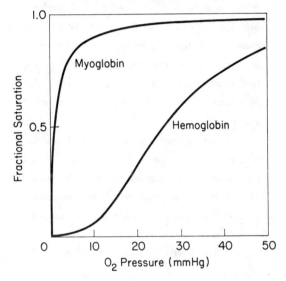

This form for the hemoglobin curve tells us the reaction is $\boxed{\text{similar to/more complex than}}$ that shown by myoglobin.

163

has instead a sigmoid form

more complex than

164

This difference arises from the circumstance that the hemoglobin molecule has four subunits, and that each of these four subunits does not react independently with oxygen. This phenomenon actually has an explanation parallel to the one given in Part II, Items 32–35, and should be reconsidered when you have studied those items.

Rhetorical question: How does it come about that the form of plotting the association (dissociation) reactions we have used in the preceding two items reappears and is used repeatedly in Part II, for enzyme catalysis?

This common ground for plotting arises from the circumstance that enzymes act by associating a substrate molecule to themselves. The greater the total number of enzyme molecules that have been converted to the enzyme-substrate complex, the more rapidly the catalytic reaction will proceed. Hence the rate will be proportional to the degree of association, and the binding curves serve also for kinetics.

Suppose myoglobin and hemoglobin were actually enzymes, and able to catalyze the conversion of O_2 into another substance, for example, 2O. If this were true, could we logically add another label on the ordinate scale of our plot of Item 3, with no change in the numbers, showing % of maximal velocity of O_2 consumption? _____

165

Conversely, bear in mind that the scale on which you plot the velocity of an enzyme reaction in Part II can logically be replaced by another scale, showing the _____ of the substrate with the enzyme molecule.*

---

*This parallelism permits interchangeable use of the several styles (the so-called *linear transformations*) developed for plotting data for enzyme kinetics and for showing binding or association processes. The longer *Enzyme Kinetics* program develops these parallel plots more extensively.

164

Yes

165

per cent association (or per cent saturation, or fractional saturation or binding; or, reversing the scale, per cent dissociation)

THE BICARBONATE-CARBONIC ACID BUFFER SYSTEM UNDER PULMONARY CONTROL

(The remaining sections, Items 166 to 193, may not prove suitable for all programs of study.)

166

In Item 74 we considered the dissociation behavior of carbonic acid. We place the apparent position of pK_1' at pH 6.1, for the reaction by which HCO_3^- and H^+ are formed.

Below the position assigned to the titration curve for this step in the dissociation is reproduced.

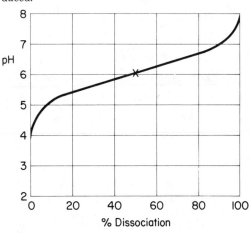

The normal pH of the plasma and the interstitial fluid is about 7.4. Draw a horizontal line across the graph to represent this pH. This pH is 1.3 units removed from the apparent pK'. From this degree of separation we would expect an ordinary buffer system to be ⟨ nearly optimal/comparatively weak ⟩ in its resistance to pH change.

167

This is, however, an extraordinary buffer system, in that one of its members, namely H_2CO_3, is unstable and in equilibrium with a gas, CO_2. Furthermore, the pressure of this gas to which the blood is exposed in the lungs can be modified by changing the rate and intensity of breathing.

Suppose we were to add to an ordinary buffer of pK 6.1, enough HCl to form as much HA as there is A^- present. The pH would then be equal to the pK'_1, namely 6.1.

But when the weak acid formed is carbonic acid,

$$HCO_3^- + H^+ \rightarrow H_2CO_3$$

it cannot remain at the new, high concentration arising from this acidification. Instead it promptly breaks down and in the living organism is excreted as CO_2 by way of the _____ .

168

If the apparent $[H_2CO_3]$ returns to its original concentration, the only change will be that we shall have approximately halved the $[HCO_3^-]$.

166

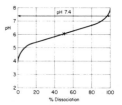

comparatively weak

167

lungs

From the Henderson–Hasselbalch equation we can see that if $\dfrac{[HCO_3^-]}{[H_2CO_3]}$ had a value such that the pH was 7.4, halving the value of that ratio will lower the pH by _____ , to _____ .

168

0.3 (Item 18)

7.1

169

Thus this system returns the pH much more nearly to the original value than would an ordinary buffer. In the living organism, it can do even better in that respect than we have so far claimed.

If the rate and intensity of respiration were to increase somewhat, the CO_2 pressure in the air in the lungs would be lowered, so that the blood would come to an even lower H_2CO_3 than it had originally. This response by the respiratory system would have the effect of lowering the pH/bringing the pH even nearer the original one .

169

bringing the pH even nearer the original one

170

In alkalosis, the $[HCO_3^-]$ of the body fluids rises rather than falls, thereby providing part of the H^+ needed in the biological deficiency:

$$H_2CO_3 \rightarrow HCO_3^- + \underline{\hspace{3cm}} .$$

But again the respiratory response controls the $[H_2CO_3]$ so as to make the pH change much larger/smaller than one would obtain with an ordinary buffer. In this case CO_2 is retained to return $[H_2CO_3]$ to about its normal value.

171

The compensation by the $HCO_3^- - H_2CO_3$ system is, however, not an ultimate one. The hydrogen-ion excess has so to speak "gone underground," but it is still present. When an acidosis is to be eliminated in an ultimate way, the lowered $[HCO_3^-]$ seen in Item 168 would have to be restored to its normal level, through the reaction,

$$H_2CO_3 \rightarrow \underline{\hspace{2cm}} + \underline{\hspace{2cm}} .$$

Here we see the H^+ excess reappearing for ultimate excretion by the kidney, as discussed in Items 186 to 191 below.

172

When the human or animal organism accumulates an excess of H^+ in acidosis, a substantial but not a major part of this H^+ is "disposed of" by the reaction we have been examining:

$$HCO_3^- + H^+ \rightarrow H_2CO_3 \rightarrow H_2O + CO_2 .$$

A larger part of the H^+ enters the various cells of the organism to be buffered there.

Even though the decrease in the $[HCO_3^-]$ of the serum does not correspond to a major part of the accumulated hydrogen ion excess, in acidosis it provides a valuable measure of that excess.

Construct a parallel statement for the situation in alkalosis. Even though _____ _____ it provides a valuable measure of that deficiency.

170

(the full equation)
$$H_2CO_3 \rightarrow HCO_3^- + H^+$$

smaller

171

(the full equation)
$$H_2CO_3 \rightarrow HCO_3^- + H^+$$

172

Even though the in-
crease in the serum
[HCO$_3^-$] does not cor-
respond to a major
part of the H$^+$ de-
ficiency in alkalosis
. . . (or equivalent
words)

BALANCING METABOLIC H$^+$ RELEASE AND FIXATION IN THE ORGANISM

173

On superficial examination our normal metabolism
seems to produce H$^+$ and to consume or fix H$^+$ so
freely as to make for a complex situation. Actually, a simple principle serves to make the subject
quite simple.

*The equation for every metabolic reaction
must be electrically balanced.* It will be apparent
when H$^+$ is needed to balance it electrically.

Take for example the sequence of reactions
known as glycolysis, totalling approximately as
follows:

$$\text{glucose} \rightarrow 2 \text{ lactate}^- + \underline{\hspace{1cm}} H^+$$

The lactate anion cannot be formed from the
neutral substance glucose, without releasing
_____ .

174

In the normal course of our metabolism, the lactate formed is ultimately oxidized to the uncharged products, CO_2 and water

$$lactate^- + \underline{\hspace{1cm}} + 3O_2 \rightarrow 3H_2O + 3CO_2$$

Accordingly, if we look at the complete catabolism of glucose,

$$glucose + 6O_2 \rightarrow 6H_2O + 6CO_2$$

we see that hydrogen ions will be neither released nor fixed by the overall process. This conclusion follows from the circumstance that *uncharged* products are formed from _____ reactants.

175

For fat oxidation we can set down similar sequences:

1) $CH_3(CH_2)_{14} COO[R] + 7O_2 \rightarrow 8\ acetate^- + 8 \underline{\hspace{1cm}}$

2) $acetate^- + \underline{\hspace{1cm}} + 2O_2 \rightarrow 2H_2O + 2CO_2$.

An esterified palmitoyl group is shown in step 1 converted to acetate and _____ ions. But in the second stage both of these intermediates are eliminated again.

173

2

H^+

174

H^+

uncharged (neutral)

175

H$^+$

H$^+$

hydrogen

176

For the catabolism of most of the neutral amino acids, and of neutral amino acid residues in typical positions in proteins, the overall situation is the same:

$$RCH(NH_3^+)COO^- \xrightarrow{\quad O_2 \quad} \tfrac{1}{2}\,urea\, +$$

carbon dioxide + water.

If we are dealing instead with the catabolism to urea, CO_2 and water, of glutamate$^-$ or arginine$^+$ in the forms these would have in foods at neutrality, as illustrated here,

$$^-O_2CCH_2\,CH_2\,CH(NH_3^+)COO^-$$

the glutamate anion

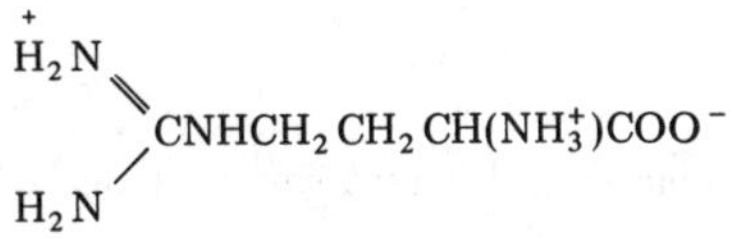

$$\begin{matrix} \overset{+}{H_2N} \\ \diagdown \\ \diagup \\ H_2N \end{matrix} CNHCH_2\,CH_2\,CH(NH_3^+)COO^-$$

the arginine cation

can effects on the H$^+$ also be neglected? _____

176

No (The total catabolism of glutamate$^-$ can be balanced only by supplying a H$^+$, that of arginine$^+$ only through the release of H$^+$.)

177

Note that we do not need to consider the detailed pathways of these sequences. As long as we set down accurately the overall result of a sequence, we will know whether or not an overall effect on the neutrality will be produced. Suppose, however, that metabolism becomes deranged in such a way that the initial, $\boxed{\text{H}^+\text{-fixing/H}^+\text{-releasing}}$ stage in the sequences of Items 174 and 175 runs far ahead of the second, H^+- _____ stage.

178

In intense muscular activity, for example, a transient lactate accumulation can occur, with an associated transient | acidosis/alkalosis | . Or when the first stage of fatty acid oxidation is caused to run far ahead of the final stage, the accumulation of acetate⁻ (in the real case, mainly of acetoacetate⁻) will be accompanied by a severe | oversupply/deficiency | of hydrogen ions. This is the situation known as *ketosis*, seen in diabetic acidosis and in other situations where a rapid, incomplete oxidation of fatty acids is stimulated.

179

An example of a different sort, but one of great biological importance, may be seen in the generation of CO_2 in our cells and its excretion via the respiratory route. During its transit through the blood stream, the CO_2 is hydrated to H_2CO_3, which dissociates to the bicarbonate ion,

$$CO_2 + H_2O \rightarrow H_2CO_3 \rightarrow HCO_3^- + \underline{\hspace{1.5cm}}$$

but these changes are reversed in the lungs,

$$HCO_3^- + H^+ \rightarrow H_2CO_3 \rightarrow H_2O + \underline{\hspace{1.5cm}}$$

so that there is no over-all effect on the neutrality.

177

H^+-releasing

fixing

178

acidosis

oversupply

179

H^+

CO_2

180

What makes this example so important is the tendency of the two stages to proceed at unbalanced rates in certain abnormal situations.

After studying the sequence in Item 179, indicate which of the following cases illustrates acidifying and which illustrates alkalinizing imbalances between these two stages:

a) A subject is hindered from excreting CO_2 by breathing into and out of a paper bag held over his face ⌐acidifying /alkalinizing⌐ .

b) A subject voluntarily breathes rapidly and deeply until the CO_2 content of his lungs and blood is brought far below the normal values ⌐acidifying/alkalinizing⌐ .

180

acidifying

alkalinizing
(These are the respiratory disturbances of the neutrality.)

181

(Continuing the same list:)

c) HCO_3^- is excreted in increased quantities into the urine, or lost in diarrhea fluid, or by the vomiting of intestinal juices ⌐acidifying/alkalinizing⌐ .

d) HCO_3^- is administered as $NaHCO_3$ by mouth ⌐acidifying/alkalinizing⌐ .

181

acidifying

182

It is a curious convention that an alkalosis produced by dosing with $NaHCO_3$ is classified as a *metabolic* alkalosis. Note that such an agent will serve in practice to combat an acidosis. For intravenous administration, an isotonic solution of sodium lactate has been convenient.

By examining Item 174, we may understand why lactate administration is alkalinizing. Write here an electrically balanced, over-all equation, in words if you like, for its complete catabolism.

alkalinizing

(These non-respiratory disturbances are generally classified with the *metabolic* disturbances illustrated in Item 177.)

183

Another important metabolic way in which net H^+ release can occur is produced experimentally when NH_4^+ is administered, e.g., in the form of NH_4Cl. Urea synthesis in the liver is stimulated, as follows:

$$2NH_4^+ + CO_2 \rightarrow \text{urea} + H_2O + \underline{\hspace{2cm}}.*$$

We see administered NH_4^+ as an | acidifying/alkalinizing | agent.

182

lactate$^-$ + H^+ + $3O_2 \rightarrow 3H_2O + 3CO_2$
(only electrical balancing was requested)

184

All the examples given so far of H^+ release or H^+ fixation in our metabolism can be pictured as stages in metabolic sequences, each of which is neither acidifying nor alkalinizing when considered as a whole.

If we consider that catabolizable organic anions are natural components of ordinary dietaries particularly in fruits and vegetables, we may classify

183

$2H^+$

acidifying

---

*We might consider this reaction as an acidifying stage of amino acid catabolism, in which case we would consider the alkalinizing stage to be,

$$RCH(NH_3^+)COO^- + H^+ \xrightarrow{O_2} NH_4^+ + \text{water} + \text{carbon dioxide}.$$

Ammonia accumulation is so injurious that any accompanying alkalinization would have to be regarded as a trivial consequence.

the oxidation of these anions as a normal acidifying/alkalinizing contribution of metabolism.

A reaction that produces net quantities of H$^+$ is normally associated with protein catabolism. This reaction is the oxidation of the sulfur of certain amino acids to sulfate. For species or populations that consume high-protein diets, this effect will dominate over the alkalinizing effect just mentioned. The reaction is illustrated for methionine:

$$CH_3 - S - CH_2\,CH_2\,CH(NH_3^+)COO^- + O_2 \rightarrow$$

$$\text{carbon dioxide} + \text{water} + \frac{1}{2}\,\text{urea} +$$

$$SO_4^= + \underline{\qquad} .$$

184

alkalinizing

2H$^+$

185

acidosis

185

For such species or populations, the threat of H$^+$ excess is much more serious and persistent than that of alkalosis. Furthermore, patients too sick to eat subsist on a high-protein fuel since their protoplasm is sacrificed for energy. Hence the tendency in disease is toward acidosis/alkalosis , especially because the renal compensatory actions to be described below are handicapped or even prevented.

ULTIMATE CORRECTION OF THE NEUTRALITY BY THE KIDNEY

186

When plasma with an excess of H^+ reaches the kidney, this organ responds by lowering the pH of the urine sharply. This action achieves little by the extra free (unbound) H^+ it eliminates, because even at the lowest pH that can be attained (about pH 4.5) the free H^+ that can be eliminated in a liter of urine would be only (check nearest estimate)

☐ about 100 milliequivalents

☐ about 10 milliequivalents

☐ less than 0.1 milliequivalent

187

The response of the kidney in acidifying the urine is effective because the urine contains solutes that come to bind the H^+ as its concentration is increased from $5 \times 10^{-8}N$ (pH 7.4) to $3.1 \times 10^{-5}N$ (pH 4.5).

The adult on an American diet ingests and excretes over a gram (roughly 40 millimoles) of inorganic phosphorus per day. From the Henderson–Hasselbalch equation and the pK′ of 6.8, we can estimate that at pH 7.4, 80% of this phosphate would be in the form $\boxed{H_2PO_4^-/HPO_4^{2-}}$; and 20% would be in the form _____ ; whereas at pH 4.5 nearly 100% would be in the form _____.*

---

*Some students may prefer to make these estimates by referring to the Figure of Response 90, rather than by use of the Henderson–Hasselbalch equation.

186

☑ less than 0.1 mEq (a liter of aqueous solution at pH 4.5 should contain 0.03 mEq of free H^+)

187

HPO_4^{2-}

$H_2PO_4^-$

$H_2PO_4^-$

188

32 (80% of 40 milli-

equivalents)

189

(the full reaction)

$HCO_3^- + H^+ \rightarrow H_2CO_3$

$\rightarrow H_2O + CO_2$

No

188

This change would mean that approximately $\boxed{100/32/4}$ milliequivalents of H^+ would be excreted in the more acidic urine, without any change in the total amount of phosphorus excreted.

189

The glomerular filtrate also contains the bicarbonate ion at a concentration similar to that of the plasma. When the urine is acidified, H^+ will be consumed or fixed by the now familiar sequence

$$HCO_3^- + H^+ \rightarrow \underline{\hspace{3cm}} \rightarrow \underline{\hspace{3cm}}.$$

Does the circumstance that the CO_2 formed cannot be retained at the high resulting P_{CO_2}, but diffuses back into the blood stream to be excreted by the lungs, decrease the effect of this reaction?

190

Another distinct phenomenon by which the kidneys excrete extra H^+ is through the synthesis and excretion of NH_4^+.* Although the kidney forms this NH_4^+ by the action of glutaminase on glutamine, the most abundant blood amino acid, and not from urea, nevertheless urinary NH_4^+ excretion occurs at the expense of urea excretion.

*Note that NH_4^+, $pK' = 9.4$, will be almost entirely in that form, and not in the form NH_3, whether the urine is at pH 4.5 or 7.8.

That is, for every gram of N excreted as NH_4^+, one gram less will be excreted as urea.

The equation of Item 183 shows that when NH_4^+ is converted to urea, H^+ must be released. This tells us that the substitution of an ammonium ion for urea N for excretion will be equivalent to the

elimination of a hydrogen ion/conservation of a hydrogen ion .

191

In alkalosis, the urine is made less acidic, possibly as alkaline as pH 7.8, and the ammonium ion excretion is decreased to essentially zero. Which of the following effects will conserve H^+ to deal with the deficiency of H^+ known as alkalosis?

☐ decrease in the amount of the urinary phosphate excreted as $H_2PO_4^-$

☐ increase in the metabolically-produced CO_2 excreted as urinary HCO_3^- rather than as exhaled CO_2

☐ decrease in the proportion of the urinary N excreted in the form of NH_4^+

190

the elimination of a hydrogen ion

191

All three contribute to the conservation of H^+.

SUMMARY

192

A weak acid is a substance that _____ _____ ; a weak base, a substance that _____ _____ ; but these reactions occur mainly when a certain characteristic $[H^+]$ has been reached. The reactivity or buffering action of the conjugate pair is distributed symmetrically about a certain point on the pH scale, namely the _____ .

192

dissociates a hydrogen ion

binds a hydrogen ion

pK'

193

The pK' has the added meaning that it is the _____ of K', the dissociation constant of the weak acid or weak base. Write the Henderson–Hasselbalch equation for

    a) a practical set of conditions _____

      _____

    b) infinite dilution _____

      _____

193

negative logarithm

a) $pH = pK' + \log \dfrac{[A^-]}{[HA]}$

b) $pH = pK + \log \dfrac{[A^-]}{[HA]}$

194

Insert into this equation the quantities for calculating the pH of the following solutions:

    acetic acid, 20 mmoles, and sodium acetate, 10 mmoles, in 100 ml water

$$pH = pK' + \log \frac{[\quad\quad]}{[\quad\quad]}$$

acetic acid, 20 mmoles, and sodium hydroxide, 10 mmoles, in 100 ml water

$$pH = pK' + \log \frac{[\qquad]}{[\qquad]}$$

ammonium chloride, 20 mmoles, and sodium hydroxide, 10 mmoles, in 100 ml water

$$pH = pK' + \log \frac{[\qquad]}{[\qquad]}$$

195

A dissolved substance bearing two dissociating groups on the same molecule shows

☐ 1. a titration behavior not distinguishable *a priori* from that expected if two dissociating substances were present

☐ 2. a single sigmoid titration curve centered midway between the two pK' values

☐ 3. an isoelectric point midway between the two pK' values

196

If we are to calculate the pH of a phosphate buffer, we decide which pK' to use in the following way:

☐ 1. If the buffer is made from H_3PO_4 and NaOH we use the first pK'; if from NaH_2PO_4 and NaOH, we use the second pK'.

☐ 2. We take into account which two of the four species H_3PO_4, $H_2PO_4^-$, HPO_4^{2-}

(or for all conditions)

$$pH = pK + \log \frac{a_{A^-}}{a_{HA}}$$

194

(The fractions should show the ratios.)

$$\frac{0.10}{0.20}, \frac{0.10}{0.10}, \frac{0.10}{0.10}$$

195

☑ 1. titration behavior not distinguishable

. . .

and PO_4^{3-} are most abundant, and then we use the pK' to which these two pertain as a conjugate pair.

196

☐ 1. (not necessarily!)

☑ 2. We take into account . . .

197

The titration behavior of larger polypeptides and proteins is in general:

   ☐ 1. Very similar to that of contained amino acids, since the titratable groups present are just a sum of all those of the amino acids.

   ☐ 2. determined largely by the dissociating groups of the amino acid sidechains; most of the α-amino and α-carboxyl groups of the contained amino acids make no contribution.

(The remaining 4 review items pertain to the biological application of Items 161–186.)

197

☑ 2. determined largely by . . .

198

The effectiveness of the HCO_3^- – H_2CO_3 buffer in the biological context depends on the normal maintenance of a rather high carbon dioxide pressure within the organism, which permits both the $[H_2CO_3]$ and the $[HCO_3^-]$ to be kept much higher than would otherwise be the case.

For a given hydrogen ion excess or deficiency, this buffer suffers a much | larger/smaller | change in pH than would occur with a fully stable weak acid-conjugate base system.

As a result of this buffering, the extra hydrogen ions are | eliminated permanently/bound transiently until they can be excreted | .

199

In the following equations for metabolic processes, introduce one or more hydrogen ions on either side of the equation to balance the equation as to electric charge:

1) $lactate^- + 3O_2$ _____ $\rightarrow 3CO_2 + 3H_2O$

_____

2) $lactic\ acid + 3O_2$ _____ $\rightarrow 3CO_2 + 3H_2O$

_____

3) $CH_3(CH_2)_{14}COO[R] + 7O_2$ _____ $\rightarrow$ $4CH_3COCH_2COO^- + 4H_2O$ _____

4) $RSH + O_2$ _____ $\rightarrow$ carbon dioxide + water + $SO_4^=$ _____

5) $2NH_4^+ + CO_2$ _____ $\rightarrow$ urea + water

_____ .

200

Suppose as an experiment you ingest a dose of NH_4Cl. You find in the urine you collect during the following 48 hours an amount of NH_4^+ about half as large as that ingested.

Does this result mean that only about one half of the ingested NH_4^+ was converted to urea?

Explain your answer in a few words. _____

_____

_____

198

smaller

bound transiently until

. . .

199

	left	right
1)	$+H^+$	$-$
2)	$-$	$-$
3)	$-$	$+4H^+$
4)	$-$	$+2H^+$
5)	$-$	$+2H^+$

200

No. The NH_4^+ appearing in the urine was formed mainly from glutamine in response to the acidifying effect of the NH_4^+ ingested, and is not the same NH_4^+ ingested. (Failure of the liver to convert any substantial amount of ingested NH_4^+ to urea would produce toxic symptoms.)

201

☐ 1. (not an ultimate elimination)

☐ 2. (Effect is negligible.)

☑ 3. Renal activity produces . . .

201

Of the following statements, which is the best exposition of the strategy used by the animal organism to eliminate in an ultimate way metabolic excesses or deficiencies of the H^+?

☐ 1. The lungs excrete more CO_2 or less CO_2, thereby eliminating more H^+ or less H^+, according to the need.

☐ 2. The kidney varies the quantity of free H^+ excreted in the urine to produce balance.

☐ 3. Renal activity produces H^+-fixing reactions, e.g., $H^+ + HPO_4^{2-} \rightarrow H_2PO_4^-$, to match the H^+-generating metabolic process, or H^+-releasing reactions to match the H^+-fixing metabolic process, according to which metabolic process predominates.

PART II

ENZYME KINETICS

INTRODUCTION

1

Suppose we want to use the rate of an enzymatic reaction to measure the concentration of an enzyme, for example in blood plasma or serum. For that purpose we should want the rate of the reaction to be limited by the amount of the enzyme/substrate present.

2

Hence, the substrate should be present in amounts just equivalent to/in generous excess of the quantity the enzyme will convert.

3

Similarly, if the enzyme requires any natural cofactor (coenzyme, cosubstrate or the like) for optimal activity, this cofactor should be present in excess/limiting amounts .

4

Before we can specify at what level the enzyme will certainly prove to be rate-limiting in the reaction used to measure the enzyme, we need to explore the effect of varying the substrate concentration over a range where this factor changes the reaction rate.

For many chemical reactions, for example the cleavage of sucrose in warm acid solution, the reaction rate continues to increase as the concentration of the sucrose is raised. In the accompanying graph, the line illustrating this behavior is line A/B/C .

1

enzyme

2

in generous excess of

3

excess

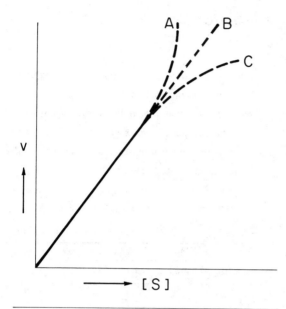

5

But as shown in the figure below, the rates of *enzymatic* reactions usually do not continue to increase with substrate concentration. At first, the rate does rise almost linearly with substrate concentration, but at higher concentrations this

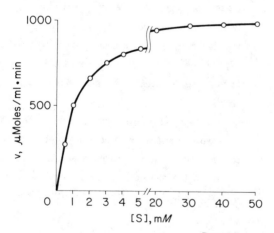

4

B

increase gradually falls off, and ultimately the rate can hardly be increased further by supplying more substrate.

Because the rate can no longer be caused to rise, we say that the substrate has *saturated* the enzymatic process. We call the rate attained the *maximal velocity*, indicated by V_{max}.

(Notice that this definition does not deny that still higher rates might be reached by some other change, for example on raising the temperature.)

In the figure the value of V_{max} appears to be about _____ .

5

1000 micromoles per
ml·min

6

a relatively very high

(Regrettably, the
terms are subjective)

6

Notice also that it is not so easy to tell from the curve of Item 5 the exact value of V_{max}. The curve is by no means an arc of a circle; it approaches the maximal velocity asymptomatically, i.e., very gradually. To attain the maximum velocity we need | only a modest/a relatively very high | concentration of the substrate.

ORIGIN OF THE RATE LIMITATION
IN ENZYMATIC REACTIONS

7

Now, what is the source of the limitation that causes the rate to stop rising steadily with the substrate concentration in Item 5? In other words, why does the substrate tend to _____ the enzymatic process?

The origin of this limitation touches on the question of what the enzyme is doing to cause the reaction to take place. It was long ago proposed that the enzyme, E, forms with the substrate, S, an intermediate *complex*, ES:

$$E + S \rightleftarrows ES$$

The product is then formed from the enzyme—substrate complex. We have written this equation to show that the formation of the enzyme—substrate complex is | reversible/irreversible | .

8

Enzymologists subsequently have shown in a number of ways that an ES complex is actually formed. Its formation may be signalled by a new absorption band, or even a color when the new absorption falls in the visual range. Slowly reacting substrates may actually be visualized in place on the enzyme molecule by the X-ray diffraction method. Hence we must regard the formation of the ES complex as (check box for the correct answer):

☐ a convenient but unsupported hypothesis
☐ a reality

7

saturate (less satisfactory, *limit*)

reversible

8

☑ a reality

9

reactive site

both

10

E, S

P

9

Some of these techniques have permitted the enzymologist to recognize a particular, limited portion of the surface of the enzyme as the *reactive site* at which the ES complex is formed. It has been possible to determine which amino acid sidechains of the protein structure contribute to the selectivity and firmness of binding the substrates at this _____ , and also to identify the amino acid sidechains that determine what catalytic effect takes place. | The first/the latter/both | of these kinds of amino acid side chains is/are therefore believed to join in forming the part of the enzyme called the reactive site.

10

Once the enzyme-substrate complex ES is formed it may react either to regenerate the original reactants, _____ and _____ , or, as shown here,

$$ES \rightarrow P + E$$

to release the product, _____ , and the original, unmodified enzyme.

11

We will now allow ourselves to consider this second reaction irreversible, by agreeing to test the initial rate before any appreciable amount of P has accumulated, so that no significant reversal can take place. Rates of enzymatic reactions measured under this condition are called *initial-rate* measurements (or sometimes *flux* measurements), and the equation can be written:

$$E + S \rightleftharpoons ES \rightarrow P + E.$$

We could instead study the initial rate of the reverse reaction of $E + P$, in which case we must make our observation before a significant amount of S/P accumulates.

12

By examining the equation for the reaction,

$$E + S \rightleftharpoons \underline{\hspace{1cm}} \rightarrow P + \underline{\hspace{1cm}}$$

we see that a molecule of the enzyme is tied up or engaged in the reaction for a definite interval of time. For that instant this molecule of enzyme is totally unavailable to another molecule of substrate. The factor that limits the rise in the rate of the reaction as illustrated in Item 5 (here repeated) is the gradually increasing proportion of the enzyme molecules tied up into the complex, until finally when all the enzyme molecules are tied into the form ES, we reach the _____ velocity.

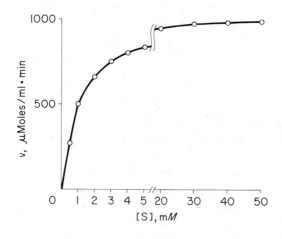

11

S

12

(the full equation)

$$E + S \rightleftharpoons ES \rightarrow P + E$$

maximal

13

The rate of the catalytic reaction is directly proportional to the concentration of ES formed:

$$v \propto [ES], \text{ or}$$

$$v = k\,[ES].$$

When all the enzyme has been converted to ES, so that we attain the maximal rate, this equation can be written:

$$v = k\,[E_{total}].$$

That is, the maximal rate is proportional to the total amount of enzyme present.*

If we set up an enzyme assay so that very nearly the maximal rate is attained, would we have the conditions for an assay for

☐ 1. the amount of enzyme present?

☐ 2. the amount of substrate present?

13

☑ 1. the amount of enzyme present.

---

*The value of the rate constant k of this equation will be high for an enzyme of high intrinsic activity, and low for a sluggish enzyme. When expressed in moles of substrate converted for mole of enzyme in one minute, it is called the *turnover number*. Enzymes are known that convert as few as six or as many as twenty million moles of substrate per mole of enzyme in one minute.

K$_m$ AND THE MICHAELIS-MENTEN EQUATION

14

To describe the effect of substrate concentration on the rate of an enzyme reaction we need one other measurement (another "parameter") besides the V_{max}. We need a measurement that positions the steeply rising portion of the curve of the preceding item. The quantity that has been selected for this purpose is the substrate concentration sufficient to produce a half-maximal rate. We will call this concentration the K_m. It is the concentration of a given substrate that produces a half-maximal rate for a given enzymatic reaction under a given set of conditions. The K_m is also called the Michaelis constant, after the physical chemists, Michaelis and Menten, who introduced it and the equation containing it. The equation describing a rectangular hyperbola has the general form:

$$v = \frac{a \cdot [S]}{b + [S]} .$$

The parameter V_{max} takes the place of a, and the parameter K_m, of b. The resulting equation (the Michaelis-Menten equation*) is written:

*For the derivation of the Michaelis-Menten equation, see the longer program, Christensen and Palmer, *Enzyme Kinetics*, Items 17–71, also Appendix A and Appendix D of that program. The derivation requires that $[ES]$ be considered approximately constant during the assay, that constancy arising from a steady state between ES formation and breakdown. We ignore here one popular but untrustworthy idea about the K_m, that it measures (in a reciprocal way) the affinity of E for S.

14

$$v = \frac{V_{max} \cdot [S]}{K_m + [S]}$$

15

In the kinetic plot below, the maximal velocity is 900 min^{-1}. Draw a horizontal line across the figure to represent this maximal rate. Draw an arrow pointing at the ordinate scale to represent the half-maximal rate of 450 min^{-1}. Then draw a vertical line pointing down to the abscissa scale to show what concentration of S will produce this half-maximal rate.

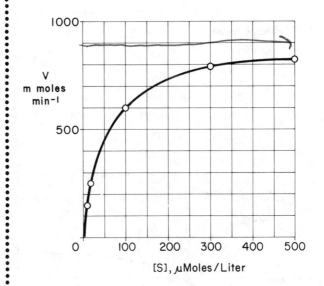

15

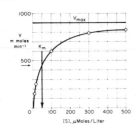

16

What is the K_m for the substrate for the enzymatic reaction pictured in the preceding item?

Suppose each of two analogous substrates, e.g., glucose and mannose, reacts with the enzyme in question to be transformed into something else. Suppose the plot we have just examined describes the reactivity of glucose. In the accompanying table, we show for comparison the rates obtained with various concentrations of glucose and mannose.

Sugar Concentration μM	v for glucose min^{-1}	v for mannose min^{-1}
10	150	82
20	256	150
100	600	450
300	770	670
500	818	750

Plot the results for mannose alongside the curve shown for glucose.

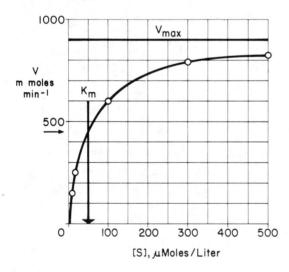

50 μmoles per liter

17

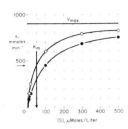

18

The completed plot suggests that mannose has a V_{max} that is

☐ very different from that for glucose.

☐ quite similar to that for glucose.

18

☑ quite similar to that for glucose.

19

Given that the V_{max} for mannose is indeed 900 mmoles/min, on the graph of Item 17, repeated below, insert a second vertical line to indicate the K_m for mannose. The result shows that it takes twice as high a concentration of mannose to produce a given velocity as it does for glucose; that is, the K_m for mannose is _____ μM, whereas that for glucose is 50 μM.

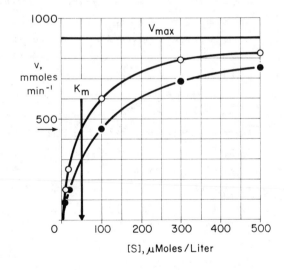

20

This example brings out a slightly confusing feature of the K_m: the more readily the substrate reacts with the enzyme, the lower will be the concentration required to half-saturate the enzyme, i.e., to produce a half-maximal velocity. In other words, when other factors are unchanged, the higher the K_m, the larger is the concentration required to produce a half-maximal rate, and the greater/smaller is the reactivity of the substrate.

21

Suppose that the concentration of a given amino acid in the liver cytoplasm tends to be 10^{-4} M. Suppose also present are two enzymes, A and B, which tend to attack the amino acid. Suppose the amino acid has a K_m of 10^{-2} M for enzyme A, and a K_m of 10^{-5} M for enzyme B. This means that the amino acid concentration falls in the range where its attack by enzyme A is slow but increases linearly with concentration, whereas enzyme B is already saturated with the amino acid.

Under these conditions we can predict that the amino acid will undergo catalytic change mainly by enzyme A/B , unless compensating differences in V_{max} are shown by the two enzymes.

Suppose some physiological event increases the concentration of this amino acid ten-fold. Will its attack by enzyme B be appreciably accelerated? _____ By enzyme A? _____

19

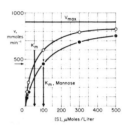

100

20

smaller

21

B

No

Yes

COMPETITION BETWEEN SUBSTRATES; COMPETITIVE INHIBITORS

22

Suppose we were to study the rate of the reaction when *both glucose and mannose* are present with the single enzyme in Items 15 to 19. Now we obtain a mixture of two enzyme–substrate complexes:

$$E + G \rightleftharpoons EG \rightarrow E + P_1$$

$$E + M \rightleftharpoons EM \rightarrow E + P_2.$$

Neither glucose nor mannose can form as much of its complex with the enzyme as it could if each sugar were present alone. Accordingly the presence of each substrate will slow the enzymatic attack on the other.

In this situation, we call mannose a *competitive* inhibitor of the enzymatic reaction with glucose, and glucose a _____ of the reaction with mannose.

23

Suppose we raise sharply the concentration of glucose. By raising the glucose level sufficiently, we can again place almost all of the enzyme in the form of its complex with glucose, and eliminate the slowing of its attack produced by the added mannose. Hence, the parasitic effect of the other sugar in trapping some of the molecules of the enzyme can be overcome. This effect brings out the meaning of the term *competitive inhibitor.* By raising sufficiently the substrate concentration, we can ultimately overcome the action of the _____ inhibitor.

24

Suppose that another substance can also combine reversibly with the enzyme at the same reactive site as the substrate:

$$E + I \rightleftharpoons EI$$

Now S is unable to bind at that site. But suppose the enzyme were unable to catalyze any modification of I. Would this substance also act as a competitive inhibitor? _____

25

Competitive inhibitors of the latter type have, in fact, proved the more valuable for artificially controlling biological processes. Such inhibitors have the advantage that they are not destroyed by the enzyme; hence their effect may be quite persistent.

22

competitive inhibitor

23

competitive

24

Yes (since it combines reversibly)

A classical case is the enzymatic dehydrogenation of succinate to form fumarate:

$$
\begin{array}{ccc}
\begin{array}{l} COO^- \\ | \\ CH_2 \\ | \\ CH_2 \\ | \\ COO^- \end{array}
&
\xrightarrow[-2H]{\text{Enzyme}}
&
\begin{array}{c} H \quad\; COO^- \\ \backslash \;\; / \\ C \\ \| \\ C \\ / \;\; \backslash \\ {}^-OOC \quad\; H \end{array}
\qquad
\begin{array}{l} COO^- \\ | \\ CH_2 \\ | \\ COO^- \end{array}
\\[2mm]
\textit{succinate} & & \textit{fumarate} \qquad\quad \textit{malonate.}
\end{array}
$$

Malonate fits the site for succinate and therefore acts as a strong competitive inhibitor of this enzymatic reaction. Can two hydrogen atoms be extracted from the malonate molecule in the same way that succinate is dehydrogenated? _____

25

No

26

Another parameter, the K_i, somewhat analogous to the K_m, is used to describe the effectiveness of a substance in inhibiting a given enzymatic reaction. The K_i is the concentration of the inhibitor required to bring the rate down to half its maximal value, given that the test is made at a relatively negligible concentration of the substrate.

The K_i has attributes similar to the K_m. It has the dimension of ⟨concentration/time⟩ , and the larger its value, the ⟨lower/higher⟩ is the concentration of the inhibitor needed to slow the enzyme reaction.*

---

*If an expanded treatment of the measurement of K_i is needed, see the longer program, Christensen and Palmer, *Enzyme Kinetics*, Items 121–125 and Appendix C.

27

Many antibiotics and other drugs produce their biological effects by acting as enzyme inhibitors. A rational approach to chemotherapy calls for the identification of strategic steps in metabolism whose modification will contribute to the effect desired. For example, an enzymatic reaction more critical to an infecting cell or virus than to the host may be inhibited.

In approaching this goal, it should be useful to compare the values of $\boxed{K_i/K_m/V_{max}}$ for a group of chemotherapeutic agents for the process under study.

28

Besides the reactive sites of enzymes, living cells present many other specific binding sites, some on their surfaces, some on their interiors. Such specific sites produce biological effects of other sorts: they may participate in transporting a substrate across a membrane and they may bind a hormone or a metabolite and thereby generate a signal to tell the cell that the hormone or the metabolite has reached an increased level. Drugs may achieve their effects by binding at any of these specific sites, and not exclusively at the reactive sites of enzymes.

Does this situation suggest that the analysis provided by enzyme kinetics as described here may be more broadly applicable to the quantitative study of drug effects? _____

26

concentration

higher

27

K_i

28

Yes

KINETICS OF TRANSPORT

29

In recent years Michaelis-Menten kinetics have been shown to describe very satisfactorily another type of biocatalysis, namely *mediated transport.** We may diagram the simplest form of the mediation of transport to show that an *active site*, with which a substrate can form a complex, is located in a membrane in such a way that it can receive the solute from one side of the membrane, and release it unchanged to the other:

$$S \ + \ | \ M \ \rightleftharpoons \ MS \ \rightleftharpoons \ M \ | \ + \ S$$

$$| \quad within \quad |$$
$$| \quad substance \ of \quad |$$
$$| \quad membrane \quad |$$

Here *M* (*Mediator*), the structure bearing the recognition site for transport, takes the place of *E*.

The transport process diagrammed above is one whose operation is symmetrical and fully ⎡ reversible/irreversible ⎤ . By introducing isotopically labeled substrate (S^*) on either side of the membrane, one can observe that its rate of appearance on the other side obeys the simple Michaelis-Menten equation, providing that one concludes the observations:

☐ before significant amounts of S^* accumulate on the opposite side.

☐ at the steady state.

---

*Occasionally one encounters the statement that transport shows "enzyme-like" kinetics. Actually, the kinetic behavior described in this program is by no means specific to enzymatic reactions, and the observation that transport shows that behavior has no force in suggesting enzyme involvement in transport. Had transport kinetics been investigated earlier, we might now be expressing surprise that *enzymes* show *transport* kinetics.

30

Suppose that the process, as stated, is symmetrical, so that the K_m and V_{max} for transport from left to right are just the same as the K_m and V_{max} for transport from right to left. At a steady state, the uncharged solute will tend to become:

☐ distributed uniformly.

☐ concentrated to one side of the membrane.

From this behavior, one might suppose that the solute had distributed itself by simple diffusion. Would that conclusion be valid in view of the correspondence of the rate to the Michaelis-Menten equation? _____ That is, should diffusion be subject to saturation by an excess of solute?

31

For this system to operate to concentrate the solute into the right-hand phase instead, V_{max} being the same in both directions, which K_m should be the larger?

☐ That for transport from left to right

☐ That for transport from right to left

(Think: At which surface will you want the higher concentration, when the two rates are equal? That surface should show the higher K_m in reacting with the solute.)

29

reversible

☑ before significant amounts of $S*$ accumulate on the opposite side

30

☑ distributed uniformly

No. (Simple diffusion follows first-order or linear kinetics, $v = k \cdot [S]$, at all concentrations.)

No. (A saturating tendency has been claimed possible under extreme conditions.)

31

☑ That for transport from right to left

ANOTHER TYPE OF INHIBITOR; METABOLIC REGULATION

32

Instead of combining at the same active site as the substrate, an enzyme inhibitor may combine at a different site on the same enzyme molecule. If that binding prevents the enzyme reaction, the inhibitor is called a non-competitive inhibitor. It may only modify reactivity to slow the over-all rate. In such cases a three-component or *ternary* complex is formed; either

$$EI + S \rightleftharpoons EIS \rightarrow EI + P$$

$$ES + I \rightleftharpoons EIS \rightarrow EI + P.$$

We find it more useful to call *I* a *modifier* rather than an inhibitor in such cases, because instead of slowing or stopping the over-all reaction, the agent might conceivably _____ it. Such modifiers appear to have important metabolic roles.

32

accelerate

33

Living organisms are able to control various steps of their metabolism by using certain key metabolites as modifiers of enzyme action and transport processes. A metabolite may control its own proper concentration in a given cell by acting as an inhibitor of an enzyme that takes part in its formation. This type of regulation is called *feedback inhibition*, because as the concentration of the product rises, its formation tends to be
| accelerated/shut off | .

34

Often in a series of enzyme reactions it is the final product that inhibits the action of the first enzyme in the series, as in the following example:

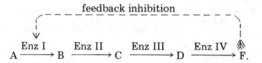

feedback inhibition

$$A \xrightarrow{\text{Enz I}} B \xrightarrow{\text{Enz II}} C \xrightarrow{\text{Enz III}} D \xrightarrow{\text{Enz IV}} F.$$

This mode of _____ inhibition is one that requires a ⎡ minimal/maximal ⎤ number of interventions by the modifier in regulating a series of reactions.

35

A number of cases are known in which the substrate itself, rather than another metabolite, slows or speeds up the action of an enzyme. These enzymes are said to show *allosteric* behavior. One of the important deviations from the rectangular hyperbola in the relation between [S] and v, namely a sigmoid form, arises as illustrated below.

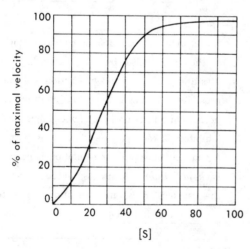

33

shut off

34

feedback

Study of this figure will show that at low substrate concentrations the rate increases rather slowly, whereas at levels of substrate about equal to K_m, the rate rises much | more/less | rapidly than is required by a rectangular hyperbola.

35

more

36

The substrate plays a double role in such cases. Frequently the molecules of such enzymes are composed of several identical subunits, each bearing an identical reactive site. The binding of the first molecule of substrate, however, modifies the properties of the other reactive sites on the same molecule of the enzyme through changes in the conformation of the whole molecule. The usual Michaelis-Menten (hyperbolic) kinetics are observed if a subunit portion of the whole enzyme molecule is separated and studied by itself. When these subunits are reassembled to form the native, intact molecule, however, the filling of one active site by changing the conformation of the molecule enhances the binding of other molecules of the substrate at the other sites, i.e., it | raises/lowers | their K_m.

36

lowers

37

Hence at low substrate levels the reaction takes place largely through catalysis by a single reactive site. Because the K_m of this site is relatively | high/low | the rate at low substrate levels will be relatively | high/low | .

38

When however, the substrate concentration is high enough so that catalysis occurs largely through the action of the reactive sites whose K_m values have been decreased, the rate will be relatively | high /low | . Thus the allosteric properties of such an enzyme permit it to respond to an elevation of the substrate level with | a slowed/an accelerated | disposition of the substrate, a behavior that tends to stabilize the concentration of the substrate.

37

high

low

38

high

an accelerated

ANOTHER WAY OF PLOTTING KINETIC RESULTS: LINEAR TRANSFORMATIONS OF THE SIMPLE MICHAELIS-MENTEN PLOT

39

So far, we have simply plotted the rate of enzymatic reactions against the substrate concentration. If we obtain a rectangular hyperbola, we say that the reaction follows Michaelis-Menten kinetics, and then the values for K_m and V_{max} provide us a complete picture of the curve. But we may see instead a deviation from the simple form, for example, under the allosteric phenomenon.

The simple plot of v against [S] has, however, two defects that we now need to deal with:

1. The eye is not very competent in recognizing any but the larger deviations from the hyperbolic form.

2. The value for V_{max} is not immediately evident from the rectangular hyperbola unless it is carried to very high substrate concentrations. Furthermore, before we can extract the value of K_m from the simple plot, we ⎡need/do not need⎦ a value for V_{max}.

39

need

40

It is a mathematical rule that if the relation between v and [S] describes a rectangular hyperbola, then a plot of 1/v against 1/[S] will yield a straight line. The same is true for a plot of v against v/[S], or a plot of [S]/v versus [S]; but we will use only the first and most common of such *linear transformations*, namely the one known as the Lineweaver-Burk plot, or as the *double-reciprocal* plot.* This plot is illustrated here for the case of glucose and mannose and the enzyme in Item 17. Do the reactions of these sugars conform to the Michaelis-Menten form? _____ What value of [S] does a value of 0.1 μM^{-1} represent for 1/[S]?

---

*If further consideration of the linear transformations is desired, see the longer program, Christensen and Palmer, *Enzyme Kinetics*, Items 78–92 and Appendix B. For the use of such equations to study binding of small molecules by proteins in general, see Items 141–159 in the same text.

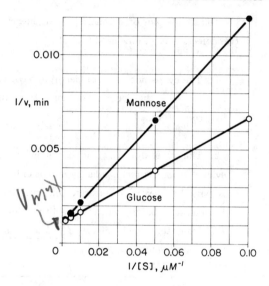

41

What value of [S] does a value of zero for 1/[S] represent? _____

42

An infinitely high substrate concentration meets the theoretical requirement for attaining V_{max}. Read the value for 1/v from the graph for this condition. _____

43

Divide this value into *one* to obtain the velocity for infinite concentration. _____ min^{-1}.
What is the value of V_{max} for both sugars? _____ min^{-1}.

40

Yes

$10 \ \mu M \ (10^{-5} \ M)$

41

$\propto$, an infinitely high concentration

42

0.001 min (or 0.0011, more precisely)

43

1000 (or 900)

1000 (or 900)

44

It can be shown algebraically that the intercept of the double-reciprocal plot with the x axis has the value of $-1/K_m$. That is, if we continue the line to negative values of $1/[S]$, we intercept the x axis at $-1/K_m$. Hence it is possible to obtain the K_m value from the graph by calculating the reciprocal of the intercept on the x axis. For glucose, the K_m appears to be _____ μM, for mannose, _____ μM.

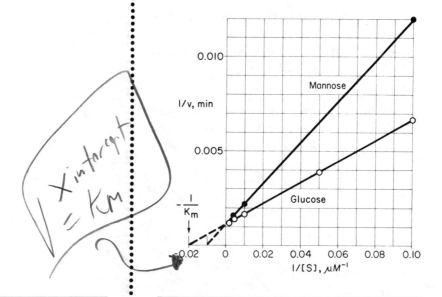

44

50

100

45

We can also use the double-reciprocal plot to determine whether the action of a given inhibitor is competitive or not. You may recall from Item 23 that the effect of a competitive inhibitor can be overcome by increasing the concentration of the substrate whereas that of a non-competitive inhibitor cannot.

This difference means that a finite quantity of a competitive inhibitor does not change the rate obtained at an infinite substrate concentration; hence the value of $\boxed{K_m/V_{max}}$ is not changed by the presence of the inhibitor.

46

Look at the two upper lines in the double-reciprocal plot shown below. Does line b or line c describe the action of a competitive inhibitor? _____ . The other line shows the presence of an inhibitor that lowers V_{max} with no effect on K_m. This effect means that the inhibitor has a $\boxed{\text{substantial/negligible}}$ tendency to occupy the reactive site of the enzyme. Accordingly it must be binding at $\boxed{\text{the reactive/another}}$ site. This type of inhibitor is known as *non-competitive*. Label the two lines to show the type of inhibition illustrated by each.

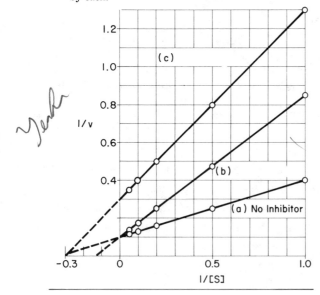

46

line b

negligible

another (Line b should
be labeled competi-
tive, line c, non-com-
petitive)

EFFECTS OF TEMPERATURE
ON REACTION RATES

47

As for other chemical reactions the rate of an enzymatically catalyzed reaction increases with an increase in temperature.

Above a certain critical temperature, however — and this critical temperature varies from enzyme to enzyme — the reaction velocity begins to decrease, and at high temperatures the enzyme becomes totally inactive. This decrease ordinarily arises from the denaturation of the enzyme.

In the figure below, mark the portion of the activity curve that describes behavior of enzymatic reactions shared with other reactions, and the portion that arises from enzyme denaturation.

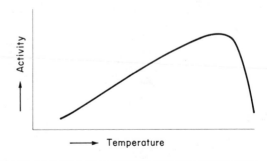

48

The increase of enzymatic activity at moderate temperatures has greater theoretical interest. The parameter most frequently used to describe the increase in rate is the $Q_{10}°$, namely the number of times by which the enzymatic activity is increased by a $10°$ rise in the temperature. If, for example, an enzyme reaction shows a V_{max} of 30 sec^{-1} at $27°$ and a V_{max} of 60 sec^{-1} at $37°$, the $Q_{10}°$ over this interval is _____ .

49

$Q_{10}°$ values of about 2 are very common for enzymatic reactions, as they are for other chemical reactions. The range observed is very wide, however. If xanthine oxidase shows a $Q_{10}°$ of 4 between $27°$ and $37°$, will it necessarily be able to carry out the oxidation of xanthine 4 times as fast at $47°$ as it does at $37°$? _____

47

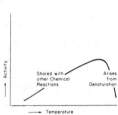

48

2

49

No (Item 47 illustrates one reason why this may not be true.)

EFFECT OF pH

50

Enzymatic reactions are sensitive to pH because:

1. certain dissociable groups on the enzyme molecule need to be in a specified state of dissociation,

2. one or more dissociable groups on the substrate may need to be in a specified state of dissociation.

The pH-activity curve for an enzyme reaction typically shows a bell-shaped curve of the sort illustrated below. We note an optimum range, and above and below this range a gradually decreasing activity.

The decreasing activity to the left of the optimum represents | an association of H^+ with/a dissociation of H^+ from | a group that loses its effectiveness with this change. In parallel, the decreasing activity to the right presumably represents _____ a group that becomes unsuitable with that change.

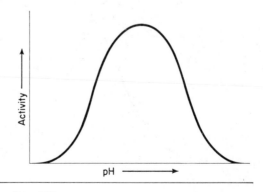

51

Can you decide intuitively whether one group is on the enzyme and one group is on the substrate? _____.

50

an association of H^+ with

a dissociation of H^+ from

52

Suppose that the figure below represents the case of an enzyme containing at the active site a glutamyl residue and a histidyl residue. Furthermore, let us assume that this enzyme will be effective only if the carboxyl group of the glutamyl residue is in its negatively charged (deprotonated) form and if the imidazole group of the histidyl residue is in its positively charged | protonated/deprotonated | form. Curves A and B are dissociation curves for these two groups. Curve A shows how the proportion of the enzyme molecules having the specified carboxyl group in its effective form increases as the pH is raised. Further

51

No

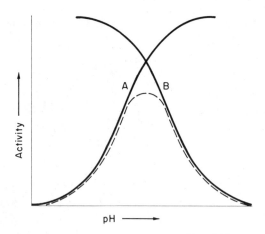

increase of the pH, however, | increases/decreases | the proportion of the imidazole group in its effective form, as shown by curve B.

52

protonated

decreases

53

indifferent

54

Yes (Consideration of the substrate should help one to decide.)

No (A dissociation by the substrate could shift the optimum.)

53

Suppose that the pH-activity curve of this enzyme reaction corresponds to the dashed-line, bell-shaped curve because of the composite of these two effects. Such a correspondence suggests that the reaction rate in this particular case is | also sensitive/indifferent | to any dissociation of the substrate in the same range. (Perhaps the substrate is a neutral molecule.)

54

In an otherwise undescribed case, is it theoretically possible that the upper limit to the pH optimum is set by a dissociable group on the enzyme and the lower limit by a dissociable group on the substrate, or vice versa?_____ Taking that answer into account, is it entirely justified to say that the pH optimum for an enzyme, e.g., pepsin, has a definite, stated value, irrespective of the substrate used? _____

55

Despite difficulties in the analysis of pH-activity curves for enzymatic reactions, their forms and the positions of the optima show important and informative differences. It is possible, for example, to measure independently the serum activity of the acid phosphatase characteristically released by a cancerous prostate gland, and that of the alkaline phosphatases released in abnormal amounts from the diseased liver or from diseased bone. In one case one may measure cleavage of a phosphate ester in an acetate buffer at pH 5, in the other in a barbital buffer at pH 8. The latter activity corresponds to the [acid/alkaline] phosphatase, and only in extreme cases is the result disturbed by accumulation of the other enzyme.

55

alkaline

IMPLICATIONS FOR PRACTICAL ENZYME ASSAYS

56

Enzyme assays are of two principal types:

1. Those in which the enzyme serves merely as a specific reagent for measuring the amount of a substrate present, e.g., how much glucose is present in blood, from the amount of a colored product form by the combined action of a glucose oxidase and peroxidase.
2. Those in which the purpose instead is to measure the amount of an enzyme present.

In the first case the enzyme is a reagent and we must be sure to have it present in [excess over the amount required to produce the conversion/a rate-limiting amount] .

56

excess over the

amount required. . .

57

generous excess

57

For the second type of assay, in which we want to assay for the amount of an enzyme in a sample, we reasoned in the opening items of this program that the enzyme must be present in amounts low enough to be rate-limiting to the reaction. Hence we concluded that the substrate and any co-factors needed must be present in ⊏ low concentrations /generous excess ⊐ .

58

Below on the standard rectangular hyperbola relating v to $[S]$, we see three regions, A, B, and C designated. We have been interested so far in this program almost exclusively in the highly curvilinear region ⊏ A/B/C ⊐ .

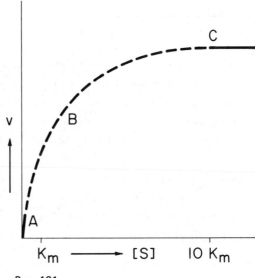

The rate of the reaction becomes largely independent of the substrate concentration in region ⟨ A/B/C ⟩ , an ideal condition if we want to obtain all the catalytic activity of which the amount of enzyme present is capable. Hence this is the region of interest in assays for ⟨ enzymatic activity/substrate concentration ⟩ .

59

Region A is the region in which the rate of conversion of substrate is essentially a linear function of substrate concentration. It is also the region in which the enzyme is in excess, so that the rate of the attack on the substrate per mole is maximal. Hence, this is the pertinent region when we are considering how much time to allow and how much enzyme to use to secure a complete conversion of the substrate in assays for ⟨ substrate concentration/enzymatic activity ⟩ .

60

Because under these conditions the rate of reaction is proportional to the substrate concentration, we can assign a constant half-time for the conversion. If we know that half of the substrate is converted in 5 minutes, then we know that half of the remainder will be converted in another 5 minutes, and so on. If we are satisfied to get 98% conversion of substrate to product, calculate how much time we should need to allow. Answer: ⟨ only 3/about 6/over 10 ⟩ half-times.

58

B

C

enzymatic activity

59

substrate concentration

60

about 6

(3 half-times, 87.5%
completion; 4 half-
times, 93.8%; 5 half-
times, 96.7%; 6 half-
times, 98.4%)

61

(All are considered
valid reasons.)

62

rate-limiting
quantities

attain (everything but
the enzyme in ex-
cess)

61

We have emphasized that it is desirable to have any enzyme used as a reagent present in generous excess. Which of the following appear to be good reasons for not setting the excess too high?

☐ The enzyme is in short supply and expensive, either to buy or to isolate.

☐ The enzyme preparation contains another enzyme that can produce an interfering reaction when too much is used — not a rare situation, considering that enzymes are prepared from tissues rich in many enzymes besides the one desired.

☐ The enzyme itself, when used in excess, can produce an interfering side reaction.

62

In assays for enzyme activity we seek to have the enzyme present in | excess/rate-limiting quantities | . Under these conditions we | attain/do not attain | the V_{max} characteristic of the substrate used.

63

Assays for enzyme activity often use two-step or coupled enzymatic reactions. In such cases one may use a second enzyme as a reagent to measure how much product the first enzyme generates as illustrated here:

$$S \xrightarrow[\text{is being measured}]{\text{Enzyme I, whose rate}} P_1$$

$$P_1 \xrightarrow[\text{not determine the result}]{\text{Enzyme II, whose rate must}} P_2 .$$

This sequence is usually chosen when P_2 is more conveniently measured than P_1. For cases resembling the one diagrammed, we want Enzyme _____ present in excess, and Enzyme _____ present in rate-limiting concentrations.

63

II (The case is covered by the principle that all reactants and co-factors should be in excess for an enzyme assay.)

I

SUMMARY

64

A rectangular hyperbola, described by the Michael-is-Menten equation, shows the increase in the velocity of an enzyme reaction with rising concentration of the | enzyme/substrate | .

64

substrate (The increase in rate with rising enzyme concentration at a fixed substrate level is also likely to be a rectangular hyperbola, but the Michaelis-Menten equation does not relate to that curve.)

65

The form of this curve is described by two parameters, as follows:

1) _____, which corresponds to the theoretical maximum approached by the curve;

2) _____, which with reference to that curve corresponds to _____ _____ _____ _____ .

65

V_{max}

K_m

the substrate concentration producing a half-maximal rate (or, the value on the abscissa corresponding to the midpoint in the rise of the curve, or an equivalent statement)

66

Suppose one substrate shows a K_m of 5 micromolar, another, of 50 micromolar. The second is a much $\boxed{\text{more/less}}$ effective substrate than the first for this enzyme reaction.

67

The reason the rates of enzyme reactions typically rise to a maximum and then increase very little with rising concentration is believed to be that ____

_____

_____

_____

68

Is it correct to conclude that raising the substrate concentration gradually decreases the chance of each substrate molecule to enter the reaction? _____ Is this behavior similar to the effect of an analog to the substrate, which also forms a complex with the reactive site? _____

69

The effect of a competitive inhibitor on the reaction rate can characteristically be overcome by adding more substrate, whereas that of non-competitive inhibitors cannot. This contrast means that a competitive inhibitor has | all/none | of its effect on V_{max} of the reaction. This feature arises from the reversibility of the binding of the competitive inhibitor | at the reactive site/at another site | .

70

A third kinetic parameter (beyond V_{max} and K_m) measures the reactivity of an inhibitor with the enzyme. This parameter, namely the _____ , resembles the K_m in that it is measured in units of | concentration/rate | .

66

less

67

the enzyme becomes fully converted to ES and therefore fully engaged by (saturated with) the substrate (or equivalent words)

68

Yes

Yes (The phenomenon of saturation with substrate and that of competitive inhibition by an analog to the substrate are in this respect inherently similar.)

69

at the reactive site

70

    K_i

    concentration

71

The regulation of metabolic processes makes use of the sensitivity of enzymes to the concentrations of substances present in the biological environment. By binding at sites other than the reactive site on enzymes, these substances act as _____ of enzyme action. Other regulatory actions occur through binding at other receptor sites present on or in the cell.

71

    modifiers (less satisfactory answer, inhibitors)

72

For a number of complex enzymes, the substrate may play a double role. When the first molecule binds as substrate, it also modifies the properties of the other reactive sites through protein-conformational changes. A sigmoid relation between the rate and substrate concentration may result, i.e., the rate at which the enzyme disposes of an accumulation of substrate is relatively ‖ high/low ‖ at high substrate concentration.

72

    high

73

Which of the following may we count as gains from converting observations of the rates of enzyme reactions at various substrate concentrations into the form of a double reciprocal plot?

    ☐ We can recognize more readily whether or not the behavior corresponds to the simplest theoretical form.

    ☐ We can extract readily the kinetic parameters, K_m and V_{max}.

    ☐ We can discriminate readily the action of competitive and non-competitive inhibitors.

74

Enzymatic reactions usually [resemble/differ from] other chemical reactions in the intensity of their response to an increase of the temperature from say $25°$ to $35°C$.

73

All three are considered valid gains.

75

Enzymatic reactions are sensitive to the pH of their aqueous environment because of changes in the degree of the H^+ dissociation

☐ of groups on the enzyme molecule.

☐ of groups on the substrate.

74

resemble

76

If an assay has as its purpose the measurement of the concentration of an enzyme in a sample, we must be sure to produce conditions under which the concentration of the [enzyme/substrate] limits the rate of the reaction. Under these conditions the reaction is proceeding at a rate [approximately equal to/far less than] V_{max}.

75

Groups on both substances may contribute.

77

If an enzyme is used instead as a reagent to measure the amount of a substrate present (or the amount of a substance produced by a preceding enzymatic reaction) we must be sure to provide this enzyme [in excess/at a rate-limiting level] .

76

enzyme

approximately equal to

77

in excess

78

(S) extreme lower

end . . . substrate . . .

(E) extreme upper

end . . . enzyme . . .

78

In relation to the usual rectangular hyperbola of the Michaelis-Menten plot, indicate on what part of the curve our observations should be made for each purpose by inserting either the letter E (assay for enzyme activity) or the letter S (assay for substrate concentration) in each of the blanks:

(   ) extreme lower end of the curve, where the rate is determined by _____ concentration

(   ) extreme upper end of the curve, where the rate is determined by _____ concentration.

PART III

USEFUL ENERGY
FROM METABOLIC
REACTIONS

* The selection of items bearing asterisks will yield a somewhat shortened sequence.

DEFINITION OF FREE ENERGY CHANGE

1

* This program considers how chemical reactions can serve to provide energy usable for the production and maintenance of life phenomena.

The possibilities are actually very limited, because neither energy made available in the form of heat nor that made available as volume expansion — the forms which drive ordinary engines — can be used to do work in the biological context. Living cells operate under almost isothermal conditions. The second law of thermodynamics states that energy supplied as heat can do no work under isothermal conditions.

These limitations leave only one rather abstract form of energy that can be generally used to drive biological processes, namely the free energy. We will represent changes in free energy by the symbol ΔG (G for Gibbs who defined the term).

Why are free energy changes so important in bioenergetics? Because in the general biological context, they represent _the only form of generally usable energy._

1

the only form of generally usable energy (or equivalent words).

2

* To determine the free energy change inherent in a reaction, we measure its tendency to proceed. We call this tendency the *spontaneity* of the reaction. Incidentally, thermodynamics defines this tendency as independent of the time required for that

Page 133

spontaneity to be realized. If we wait long enough, in which direction does the reaction tend to go, and how strongly?

The free energy change associated with a chemical reaction is not determined merely by the concentration of the reactants corresponding to equilibrium. Rather, it is determined by the relation of the actual concentrations to these equilibrium concentrations.

For the reaction, A $\rightleftharpoons$ B, the equilibrium constant is $[B]_{eq}/[A]_{eq}$, the ratio of the concentration of the product to that of the reactant at equilibrium. We compare the ratio $[B]/[A]$ existing under any given situation with that equilibrium ratio, as follows:

$$\Delta G = -RT \ln \frac{[B]_{eq}}{[A]_{eq}} + RT \ln \frac{[B]}{[A]}.$$

This equation shows that the further the ratio of the actual concentrations departs from their ratio at equilibrium, the larger/smaller will be the free energy change, whether positive or negative, per mole of reaction.

3

* This equation loses its last term when the ratio in that term is equal to *one*, because the logarithm of *one* (to whatever base) is *zero*. For that special situation, our equation simplifies to:

$$\Delta G = \underline{\hspace{3cm}}.$$

2

larger

3

$$\Delta G = -RT \ln \frac{[B]_{eq}}{[A]_{eq}}$$

(Only true when

[B]/[A] = *one*)

4

$$\frac{[B] \cdot [R]}{[A] \cdot [Q]} = 1$$

4

* We obtain the condition to which this simpler equation applies by setting each concentration at one, usually one molar, although for some reactants such as the solvent water or a metal, we set the pure substance as the standard concentration equal to one. Under these conditions, [B]/[A] = one, or for the more complex reaction

$$A + Q \rightarrow B + R, \frac{[B] \cdot [R]}{[\ ] \cdot [\ ]} = 1.$$

(Fill in the two blanks.) This situation is described as the *standard condition.*

5

* (Note that in the equation of Item 3, we are still effectively comparing the ratio [B]/[A] with the value that ratio takes at equilibrium, but under the condition that makes one of these ratios zero, so that the term containing that ratio disappears.)

The arbitrary convention then is to record in tables the value of ΔG obtained at unit concentrations of each component, and then to use the equation of Item 2 to calculate ΔG for any other set of concentrations. These tabulated values of ΔG for _standard_ conditions[1] are known as $\Delta G°$. The equation of Item 3 has already defined $\Delta G°$, thus:

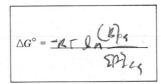

[1] In biochemistry we often define the standard concentration of H^+ as $10^{-7}N$, corresponding to pH 7. We then write $\Delta G'$ or $\Delta G°'$, rather than $\Delta G°$.

6

* The usual equation for the free energy change shows ΔG° substituted for $-\mathrm{RT}\ln K_{eq}$:

$$\Delta G = \Delta G^\circ + \mathrm{RT}\ln \frac{[B]}{[A]} \, .$$

Although this equation looks simpler, it should not cause us to lose sight of what the equation is really measuring, namely how far the actual concentration ratio departs from the value that ratio will attain at _____ .

Trap: Does ΔG° state the free energy change for a reaction taking place at equilibrium? _____

7

* In place of $\mathrm{RT}\ln\dfrac{[B]}{[A]}$ we can write $2.303\ \mathrm{RT}\log$

$\dfrac{[B]}{[A]}$, thus converting the Napierian logarithm to the logarithm to the base 10. Next, we can insert the values for the gas constant R and the absolute temperature, to obtain a factor useful for calculating ΔG° in cal./mole at 25°.

$$\Delta G^\circ = -(2.303 \cdot 1.982 \cdot 298)\log K_{eq}$$

$$\Delta G^\circ = \underline{\hspace{3cm}} \log K_{eq}, \text{cal./mole}$$

Suppose equilibrium for the reaction $A \to B$ occurs when B attains 10 times the concentration of A. Hence $[B]_{eq}/[A]_{eq} = 10$, and $\log [B]_{eq}/[A]_{eq} = $ _____ . Hence $\Delta G^\circ = $ _____ cal./mole.

5

standard

$$\Delta G^\circ = -\mathrm{RT}\ln \frac{[B]_{eq}}{[A]_{eq}}$$

6

equilibrium

No (rather, under standard conditions, namely the only practical condition under which the final term becomes zero. We use the value of K_{eq} to discover how much energy is available at $1M$ concentrations. Reread Items 5 and 6 if this query exposed confusion).

(Note that negative values for ΔG mean that the system loses energy; hence free energy is made available by the progress of the reaction.)

7

-1360

one

-1360

8

* Suppose we actually set the concentrations of A and B at equilibrium:

$$\Delta G = -1360 + 1360 \log \frac{10}{1}$$

$$\Delta G = \underline{\hspace{3cm}}$$

This means that no energy is available from the reaction at equilibrium. A student might reply, "of course the reaction will provide no energy at equilibrium, because it can proceed in neither direction." By saying that, he could be missing the main point. Suppose we arrange matters so that the reaction does proceed very close to equilibrium. For example, one stage in a sequence may provide A, and another stage may remove B, just as rapidly as A is converted to B. The free energy change would prove very small for the conversion of a mole of A into a mole of B. What thermodynamics tells us is that at equilibrium ΔG is zero calories per _____ , and not merely that it is zero calories for zero moles of reaction.[2]

---

[2] This point becomes very important when we are to estimate the efficiency of sequential or linked reactions: the more closely the linkage permits each to proceed at equilibrium, the greater the efficiency of the linkage.

9

* If we were to mix A at $1M$ with B at $0.1M$

$$\Delta G = \Delta G° + 1360 \log \frac{0.1}{1}$$

$\Delta G =$ _____ cal./mole

Now more energy is made available per mole of reaction than under the standard condition because the reaction is proceeding at a point further from equilibrium. Note that for this measurement, we should not let the ratio 0.1/1 change very much during the measurement.

Suppose instead we set [B] at $1M$ and [A] at $0.01M$. Calculate the free energy change for the same reaction.

$\Delta G =$ _____

10

* A positive value for ΔG means that free energy must be ┌ released/taken up ┐ during the course of the reaction. Can the reaction proceed in the direction A → B in the biological context? _____
(Note that for the direction B → A, the same free energy change would apply, but with the sign changed.)

8

zero

mole

9

$\Delta G = -2720$ cal./mole
$\Delta G = -1360 +$

$$1360 \log \frac{1}{0.01}$$

$= -1360 + (2 \cdot 1360)$
$= +1360$ cal./mole

10

taken up

No

THE MEANING OF COUPLING

11

* This result brings us to an emphatic paradox: Strictly speaking, within the ordinary biological context, *reactions accompanied by a positive free energy change cannot take place.*

We write many equations for biochemical reactions for conditions such that the free energy change is positive. We know that ester and amide bonds (including peptide links) are formed even though their formation involves a gain in free energy. We know that solutes are propelled against concentration gradients, even though a free energy gain is inherent in that effect. All these processes we call *endergonic,* in contrast with *exergonic* or *spontaneous* processes. In other words, they are *non-spontaneous,* and strictly speaking should not occur.

Consider the case of Item 9, where the reaction A → B is supposed to occur even when ΔG = +1360 cal./mole. It won't. Suppose that simultaneously a simple hydrolase is splitting ATP to ADP + P_i in the same solution, under conditions such that ΔG for this hydrolysis is − 7300 cal./mole. Trap: Make no unstated assumption! Do you see any reason why this second reaction should permit the reaction A → B to proceed, even though it has a positive ΔG? _____

12

✳ Biochemists sometimes get so enthusiastic in calculating that ATP breakdown releases enough free energy to "drive" an endergonic process, that they forget to stress sufficiently that an event called *energetic coupling* (or sometimes, *energy transduction*) is necessary, if one process is to drive another.

What is really required under the concept of *energetic coupling* is that our endergonic process be tied into a larger process, so that it is part of it. Furthermore, that larger process must be one that can really occur, i.e., a process for which a $\boxed{\text{gain/loss}}$ of free energy takes place.

11

No. (You shouldn't unless you've received a wrong impression earlier. At the risk of frustrating you, you should be stopped abruptly if you take coupling for granted.)

13

✳ Gregorio Weber puts the matter in this provocative way:

> Every chemical compound generated in metabolism is the result of a reaction which runs toward thermodynamic equilibrium with complete independence of any other reaction occurring at the same time.

The problem is quite parallel if a chemical reaction is to drive the transport of a solute against a concentration gradient. We have seen that a net movement of a solute from phase 1 to phase 2 cannot occur if a gain of free energy must accompany it. To paraphrase Weber's dictum:

> Every material flow in the living organism, fully described, runs toward thermodynamic equilibrium with complete independence of any other material flow and the progress of any chemical reaction occurring at the same time.

12

loss

This statement sounds as though we are denying that active transport can exist. In one narrow sense, it really can't. To allow the concept of active transport to survive, we must (choose the better completion):

☐ 1. allow that the progress of a chemical reaction or another material flow can become part of a total spontaneous process that includes the "uphill" movement of the solute.

☐ 2. assume a source of external energy not arising from concentration changes, i.e., other than the free energy.

☐ 3. deny our paraphrase of Weber's statement in this Item.

13

☑ 1. allow that . . . can become part of . . .

☐ 2. (Only in special cases can radiant energy be trapped to drive transport, and even then through chemical coupling.)

☐ 3. (We could insist on a definition of active transport incompatible with the paraphrase. Active transport so-defined could not exist.)

14

* The simplest definitions of *coupling* say merely that the two reactions to be linked must share at least one reactant. In the sequence,

$$\begin{array}{c} 1 \\ A \rightarrow B + C \\ \downarrow 2 \\ D \end{array}$$

the spontaneous progress of the first reaction can raise the concentration of B, and thus make the value of ΔG for the second reaction more negative; indeed it could change ΔG_2 from a zero value to a negative value. We sometimes say for this case that the first reaction "*pushes*" the second reaction.

Instead, the spontaneous progress of the second reaction could lower the concentration of B, to make ΔG for the first reaction more negative. We may say then that the second reaction is "pulling" the first. The two reactions of this item are, we may say, *sequentially linked.* The energy of one reaction is conserved to produce the other, although no further energy storage takes place.

Does this behavior violate the principle that all biological processes, when fully described, will be seen proceeding toward equilibrium? _____

15

* We can easily set up a similar sequential relation between a chemical reaction and the migration of a substance across a membrane. In the sequences illustrated here,

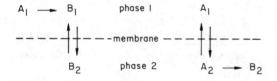

the spontaneous conversion of A into B, whether in phase 1 (left) or in phase 2 (right) can cause increased migration to occur into phase 2, so that more B accumulates there. (The subscripts 1 and 2 merely indicate the phase in which A or B is situated.)

We may say that if the chemical reaction occurs in phase 1, it "pushes" the transport; if it occurs in phase 2, it "_____" the transport.

If instead the transport is seen as the more spontaneous process, either it removes some of the

---

14

No. (As long as we take the actual concentration of B into account, we find each reaction proceeding toward equilibrium without any additional influence from the other reaction.)

product of the reaction (left), or it provides more reactant for it (right). In these cases we may see the transport as "pushing" the chemical reaction (right-hand sequence), or as "pulling" the reaction (_____ -hand sequence). Does the term *sequentially linked* apply to the relation between the reaction and the transport in this item?

_____

15

pulls

left

Yes

16

* These effects by no means correspond, however, to what we mean by *active transport*, or what we mean when we say *a chemical reaction drives an uphill transport*. Instead, by those words we mean unequivocally that the chemical reaction causes the solute to move against its concentration gradient (more precisely, for those familiar with that term, against the gradient of its *total chemical potential*).

Sequential linkage will not produce that effect. It can be achieved only by fusing the chemical reaction (or sequence of reactions) with the transport process; thereby we achieve a total chemical reaction that releases a product molecule into a phase other than the phase from which it received either that molecule or a precursor molecule:

$$A + X_1 \rightarrow B + X_2.$$

Does this fusion (coupling) of $A \rightarrow B$ with $X_1 \rightarrow X_2$ bring uphill transport within the principle that all biological processes, fully described, proceed toward equilibrium? _____

17

* By studying this distinction, which is strictly observed in the transport field, we can better understand what we want to mean by the term, *energetic coupling between chemical reactions.* Let us see how we might extend this distinction to a purely chemical coupling, and thereby define what we see as the essential nature of coupling.

First, let us reexamine our two sequentially linked chemical reactions:

	$\Delta G°$, cal./mole	ΔG, cal./mole
1) $A \rightleftharpoons B + C$	+5000	-500
2) $B \rightleftharpoons D$	-8000	[ ]
Over-all, $A \rightleftharpoons C + D$	-3000	-3000

Suppose the progress of the second reaction is able to bring the concentration of the intermediate, B, low enough, despite the continued progress of the first reaction, so that ΔG for the first reaction becomes negative as shown. Supply the missing value for ΔG for the second reaction. Under these conditions, the sequence | becomes possible /remains impossible | .

16

Yes. (Presumably all participating species are included in the equation. If so, the reaction will proceed toward equilibrium independently of other reactions and transports.)

18

* While we are examining this phenomenon, however, we should avoid exaggerating what it can accomplish. Given that [C] is held constant, the ratio of [B]/[A] would need to be enormously lowered to obtain the indicated spontaneity of reaction 1:

$$-500 = 5000 + 1360 \log [B]/[A]$$
$$\log [A]/[B] = 4.03$$
$$[A] > 10,000 [B]$$

17

-2500

becomes possible

Given that [A] is $10^{-2} M$, the concentration of B would have to be brought below $\boxed{10^{-4} M/10^{-5} M /10^{-6} M}$ to make reaction 1 spontaneous.

Consider carefully: Will the circumstance that the sum of ΔG_1° and ΔG_2° is negative in this example ensure that we will have a biologically useful linked sequence?

18

$10^{-6} M$

No. (Unless the cell can manage and tolerate that [A] be enormously greater than [B], linkage will be biologically impractical.)

19

is not

linkage

cannot

19

* For contrast, consider these two chemical reactions:

$$A \rightarrow B \qquad Q \rightarrow R.$$

They have no components in common. Now it $\boxed{\text{is/is not}}$ possible for one of them to "push" or "pull" the other; we have excluded *sequential* _____ between them. We will specify that under the real conditions, the second reaction is endergonic, which means that, *per se*, it really $\boxed{\text{can/cannot}}$ proceed in the direction $Q \rightarrow R$.

20

* In order for the first reaction to drive the second, the two reactions need somehow to be fused into one reaction

$$A + Q \rightarrow B + R$$

for which ΔG is negative, or into a single reaction sequence for which under the real conditions ΔG is negative for $\boxed{\text{the over-all process/every step}}$.

Because this fused process converts Q into R at concentrations of each such that this conversion *per se* increases the free energy, we have a situation parallel to that of active transport, where X is caused to move into phase 2 against a concentration gradient. We may say that we have achieved an energetic coupling in every sense.

We have a contrast then between two important concepts:

1. *direct energetic coupling,* parallel to energization of active transport

2. *sequential* _____

21

* Suppose our generalized reactions $A \rightarrow B$ and $Q \rightarrow R$ are really these two:

1. $ATP + H_2O \rightarrow AMP + PP_i$, $\Delta G^\circ = -8000$ cal./mole
2. $M + N \rightarrow MN$, $\Delta G^\circ = +3500$ cal./mole.

If these two equations represent a complete description of what's going on, the second reaction can indeed/cannot proceed.

But suppose the presence of an enzyme system (a *ligase*) channels the actual sequence through these two steps:

1a. $ATP + M \rightarrow AMP\text{-}M + PP_i$, $\Delta G = -2000$ cal.

2a. $AMP\text{-}M + N \xrightarrow{H_2O} AMP + MN$, $\Delta G = -2500$ cal.

20

every step (For the ΔG° values, it may be sufficient that the sum be negative; but each of values applying for ΔG under the real conditions must, as Weber emphasized, be negative.)

linkage

Note that each of the two-stage sequences, as written above, summates to the same over-all effect, namely

$$\text{ATP} + \text{M} + \text{N} \xrightarrow{\text{H}_2\text{O}} \text{AMP} + \text{PP}_i + \text{MN}.$$

Does the over-all reaction become possible along the second pathway? _____ Here we have in every sense a direct *coupling* between reactions 1 and 2. Note that sequential linkage now also occurs, as a detail of mechanism, between reactions ⟦ 1 and 2/1a and 2a ⟧ . The essential feature of the coupling, however, is that within this mechanism ATP hydrolysis presumably cannot occur without MN synthesis.

21

cannot

Yes

1a and 2a

22

The formation of PP_i and AMP in such coupled reactions has an important energetic advantage over the formation of P_i and ADP in certain other coupled reactions, in that the living cell must protect its supply and maintain a substantial concentration of this nucleotide ADP, whereas PP_i is sacrificed and its concentration kept very low through the action of pyrophosphatase:

$$\text{PP}_i + \text{H}_2\text{O} \rightarrow 2\text{P}_i.$$

Even though the two routes of ATP cleavage show rather similar values for $\Delta G°$ (-7300 to -8000 cal./mole), we may therefore expect the cleavage to ⟦ $\text{ADP} + \text{P}_i/\text{AMP} + \text{PP}_i$ ⟧ to produce much the greater change in G under the actual conditions.

23

22

We need to make in this item one more approach to the idea of a chemical reaction driving an uphill transport, as in Item 16, namely the simpler case in which one transport drives another. The figure shows a receptor site facing first one side and then another of the plasma membrane. This site is filled by the binding of both Na^+ and a solute X, and vacated by the dissociation of both together. Between the binding and the dissociation step, the site may reorientate from one surface of the membrane to the other.[3]

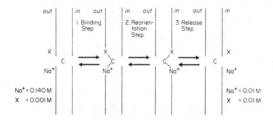

A separate process in the meantime maintains a gradient of Na^+ across the plasma membrane, with high Na^+ on the outside and low Na^+ inside.

Through the coupling illustrated here between Na^+ and X migration (a process called *cotransport*) we obtain a total, spontaneous process,

$$Na^+_{out} + X_{out} \rightarrow Na^+_{in} + X_{in}$$

so that the "downhill" movement of Na^+ can be said to drive the "uphill" movement of X. Note, the figure shows that a gradient of X has been generated.

[3] If occasionally only one of the two dissociates during exposure at a surface of the membrane, the coupling will be *loose* rather than tight.

$AMP + PP_i$

(A far lower value will be maintained for the fraction $\dfrac{[AMP] \cdot [PP_i]}{[ATP]}$ than for $\dfrac{[ADP] \cdot [P_i]}{[ATP]}$.)

THE MEANING OF COUPLING

To interpret the behavior as active (or uphill) transport, we must arbitrarily consider the migration of ⟨ X occurring alone/Na$^+$ occurring alone/both taking place together ⟩, despite the circumstance that the real process is a cotransport with a ⟨ gain/loss ⟩ in free energy.

If two separate structures, C and C′, mediated the passage of Na$^+$ and X across the membrane, no direct coupling would occur, and their flows would be ⟨ dependent/independent ⟩ .

23

X occurring alone

loss

independent

24

We have already set up the condition whereby a chemical reaction is to drive an uphill transport, namely that the exergonic reaction, $A \to B$, must be fused to the uphill transport, $X_1 \to X_2$, thus:

$$A + X_1 \to B + X_2 .$$

Thermodynamics requires that the ratio $[X_2]/[X_1]$ produced by the reaction $A \to B$ not exceed a value representing all the free energy available from that reaction under the real conditions. In molecular events one expects that theoretical limitation be expressed through a gradual acceleration of the reverse reaction

$$B + X_2 \to \underline{\hspace{3cm}}$$

as equilibrium is approached. Indeed if we were to add a good deal of X_2 to the system, we could expect its down-gradient movement to cause a synthesis of _____ .

Next page 154, Item 29

29

We were able to sense that the equilibrium point for the cotransport of Na^+ and X in Item 23

$$Na^+_{out} + X_{out} \rightleftharpoons Na^+_{in} + X_{in}$$

leaves us with a considerable gradient of Na^+, opposed by an equal gradient of X if there is no transmembrane potential. The free energy that ordinarily is released when any Na^+ moves down its gradient is captured and stored by the generation, mole for mole, of the gradient of X.

In the same way the equilibrium point in the chemical reaction of Item 21

$$ATP + M + N \xrightarrow{H_2O} AMP + PP_i + MN$$

leaves us with a $\boxed{\text{higher/lower}}$ ratio of ATP to AMP, at a given pyrophosphate level, than we would observe if the cleavage of ATP were to occur by a simple, straight-forward hydrolysis. The free energy that ordinarily is released when ATP is hydrolyzed is captured and stored in the chemical architecture of the product _____ .

30

In the real situation one would not expect to find the reaction proceeding at a thermodynamic equilibrium. Instead we expect some inefficiency in the energy transfer. Nonetheless, we may expect the living cell to contain ATP at a $\boxed{\text{higher/lower}}$ steady-state concentration with respect to its various hydrolytic products (ADP, P_i, AMP, PP_i)

$2\mathcal{4}$

$A + \mathcal{X}_i$

A

29

higher

MN (PP_i is not seen as the useful product here, and is usually further hydrolyzed.)

than would be expected if its hydrolysis were not restrained by fusion or coupling to various endergonic syntheses or transports.

When we accept that relation, we already have foreseen that an artificial raising of the concentration of MN with respect to M and N | can/cannot | cause the synthesis of ATP.

30

higher

can

31

* We encounter in another way the principle that for any biochemical reaction to occur, it must, in its actual pathway, avoid an endergonic stage.

In many cases thermodynamically spontaneous chemical reactions move so slowly toward equilibrium that for all practical purposes they do not take place. An example is the combination of H_2 and O_2 when the two gases are mixed. The reason such a reaction largely fails to occur is that one stage of the only pathway open to the reaction requires an energy input, which is spoken of as the *energy of activation*. When the barrier represented by the energy of activation is so high that almost

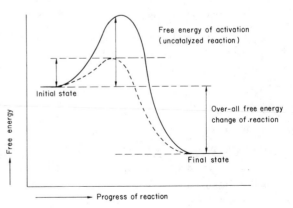

Page 155

none of the reactant molecules show sufficient kinetic energy to surmount it, no reaction occurs, even though the over-all reaction may be described as spontaneous.

A feasible pathway (dotted line) may still involve a definite energy of activation, but one for which the barrier is low enough so that many of the molecules show sufficient kinetic energy to react.

What enzymes and other catalytic systems do is to:

☐ 1. make the over-all process even more spontaneous.

☐ 2. supply the energy of activation.

☐ 3. provide an alternative pathway avoiding the "high energy barrier."

32

Perhaps the largest paradox of the discussion of bioenergetics runs as follows:

1. We speak of how much useful energy a reaction can "release," that is, how large a negative value for ΔG it shows.

2. But once that energy is actually and literally released, it ceases to be able to do work, i.e., it ceases to be useful energy and appears as heat.

When processes are coupled together, our strategy, for the sake of efficiency, is to minimize the actual release of energy by matching the exergonic process with an energy-using one, such that | as much/as little | as possible of the free energy of the first reaction can be observed as heat and high spontaneity.

31

☑ 3. provide an alternative pathway . . .

32

as little

33

Furthermore, we need to restrain the spontaneity of exergonic reactions so that as much as possible of the theoretically available free energy is not actually released but retained in the formation of another product, or of a gradient.

A cyclist knows that he does little work unless the impulse he applies to the pedal is restrained by a resistance against which he must work. By the same token, the nearer to equilibrium a linked sequence proceeds, the $\boxed{\text{better/worse}}$ it conserves the free energy inherent in its progress and the $\boxed{\text{more/less}}$ of that energy is released and dissipated as heat. It could of course proceed too slowly if too near equilibrium.

33

better

less

34

* What we quite properly do in discussing these processes is to dissect artificially an over-all process like this one,

$$ATP + M + N \xrightarrow{+H_2O} AMP + PP_i + MN$$

into two formal stages,

$$ATP \xrightarrow{H_2O} AMP + PP_i, \Delta G^\circ = -8000 \text{ cal./mole}$$
$$M + N \rightarrow MN, \Delta G^\circ = +3500 \text{ cal./mole}$$

and then we note with satisfaction that the first can provide plenty of energy to drive the second, even if concentrations depart far from the standard one molar levels.

But if the artificial separation shown here really described the situation, $\boxed{\text{less than half /none}}$ of the energy of the first reaction would be conserved by the second.

OXIDATION-REDUCTION REACTIONS AND POTENTIALS; OXIDATIVE PHOSPHORYLATION

35

There is one class of reactions for which we have an especially direct way of measuring how much useful work they can do, and also how much of the free energy they are capable of releasing is conserved. These are the oxidation-reduction reactions. If we can cause these reactions to take place in an electrical cell, we can direct the electron flow from one part of the reaction to the other through an electrical conductor, i.e., a wire. For the direction in which the electron flow occurs spontaneously, we can measure how much work it can do:

☐ 1. by seeing how large an opposed voltage is just sufficient to stop the flow.

☐ 2. by noting how much current must flow to produce one gram-equivalent of a reaction.

35

☑ 1. by seeing how
large . . .

36

Here we visualize an electrical cell for producing the reaction

$$Cu^{2+} + Zn \rightarrow \underline{\hspace{1cm}} + \underline{\hspace{1cm}}$$

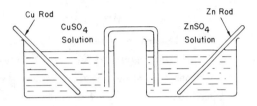

Cu Rod

CuSO$_4$
Solution

Zn Rod

ZnSO$_4$
Solution

(The reader may recall from earlier experience that the reaction proceeds spontaneously in the direction written. Do not be concerned that no metallic conductor has yet been provided between the electrodes.)

36

(the full equation)
$$Cu^{2+} + Zn \rightarrow Cu + Zn^{2+}$$

37

Because zinc metal tends, in this case, to become zinc ion, and copper ion to become copper metal, we will note that one electrode tends to become depleted, the other electrode to become enriched with electrons, relatively speaking.

Mark each of the electrodes plus or minus to indicate the states of charge arising from these tendencies of electrons to pass from the solution to the electrode, or *vice versa*.

Page 159

Each of the vessels sketched is called a *half-cell*, because both are needed to constitute a complete electrical cell. The two partial reactions:

$$Cu^{2+} + 2\epsilon^- \rightleftharpoons Cu$$
$$Zn \rightleftharpoons Zn^{2+} + 2\epsilon^-$$

are called half-cell reactions. A full oxidation-reduction reaction results because one of them uses the electrons released by the other. What needs to be added to the sketch to permit that to happen?

_____

_____

38

Note, if we did not keep the contents of the two half-cells apart, except for allowing minimal migration of ions through the U-shaped salt-bridge shown at the center, then the cupric ion would attack the zinc electrode spontaneously and uncontrollably. We restrain that attack by forcing the electron transfer to pass through the conductor. To approach the maximal work the total reaction can perform, we should:

☐ 1. not obstruct the electron flow through the wire.

☐ 2. restrain the flow further by channeling it through a device that does work.

37

The electrode at the left should be marked as plus, the other as minus.

A wire or other conductor, from one electrode to the other.

38

☑ 2. restrain the flow further by channeling it through a device that does work.

39

We can convert the oxidation-reduction potential observed (which corresponds to the voltage required barely to stop the flow through the conductor) into the maximal number of calories of work that the reaction can produce, i.e., into the free energy made available.

We define a *standard electrode potential, E°,* much as we did ΔG°, as the electrode potential observed under a standard state, and as in the case of the preceding 3 items, with _____ and _____ both at $1M$ concentrations.

We then calculate the electrode potential for any other set of concentrations by the Nernst equation:

$$E = E^{\circ} - \frac{2.303\ RT}{nF} \log \frac{[Zn^{2+}]}{[Cu^{2+}]}.$$

This equation corresponds precisely to our free energy equation, given that

$$\Delta G = -nFE$$

where n = the number of electrons involved in the reaction, F = Faraday's constant, and E = the observed _____ .

39

Cu^{2+}, Zn^{2+}
(either order)

electrode potential

40

The important thing to remember at this point is not the Nernst equation for its own sake, but that E° measures the tendency of the reaction to go to completion much as ΔG° does. Therefore, we have an especially simple way of getting that information for oxidation-reduction reactions, namely by measuring the _____ against which electrons can be caused to flow by the reaction.

41

Just as we preferred above to dissect formally the reaction,

$$ATP + M + N \xrightarrow{+H_2O} AMP + PP_i + MN$$

into two partial reactions, one exergonic and the other endergonic, and then to establish the value of $\Delta G°$ for each partial reaction, we also separate the two half-cell reactions, in this case both formally *and physically*, and establish the half-cell electrode potential contributed by each:

$$Cu^{2+} + 2\epsilon^- \rightleftharpoons Cu, E_{½}° = +0.337 \text{ volts}$$
$$Zn^{2+} + 2\epsilon^- \rightleftharpoons Zn, E_{½}° = -0.763 \text{ volts}.$$

Note, however, that by convention we write all such reactions in the direction of reduction. Draw a long arrow through each equation to remind yourself in which direction each half-cell reaction tends to occur spontaneously in the electric cell under discussion.

42

Now calculate the total potential of the electrical cell, taking into account that we must reverse the sign for the half-cell reaction that actually proceeds

---

40

potential

(voltage)

41

$Cu^{2+} + 2e \rightleftharpoons Cu \rightarrow$

$\leftarrow Zn^{2+} + 2e \rightleftharpoons Zn$

in the oxidative direction. The electrical cell can yield a potential of _____ volts. We confirm that the total reaction is $\boxed{\text{endergonic /spontaneous}}$.

42

+0.337 volts

+0.763 volts

$\overline{\text{E}^\circ = +1.100 \text{ volts}}$

(Note the sign of E° is opposite to that of ΔG°.)

spontaneous

43

☑ 2. a cell that will generate a relatively ... (If we set the reduction of copper to producing the oxidation of a metal nearer copper in the electrochemical series, we would conserve more energy within the cell itself.)

43

When we select the oxidation of zinc metal to Zn^{2+} as a reaction to be driven by the reduction of Cu^{2+} to copper metal, we illustrate:

☐ 1. a high degree of conservation of the free energy released by the reduction of Cu^{2+}.

☐ 2. an electrochemical cell that will generate a relatively high voltage, the energy to be conserved by setting it to do work outside the cell.

44

Almost any textbook of biochemistry will contain long tables of the values for the standard electrode potentials of familiar half-cell reactions of biochemical interest. Usually the values are for pH 7.0 rather than for $1N\,H^+$. We then write $E'_{1/2}$ or $E^{\circ\prime}_{1/2}$ instead of $E^\circ_{1/2}$.

We can examine here an abbreviated list.

	$E^{\circ\prime}_{\frac{1}{2}}$
O_2/H_2O	+0.815
Cytochrome oxidase (Fe^{3+}/Fe^{2+})	+0.771
Cytochrome a (Fe^{3+}/Fe^{2+})	+0.29
Cytochrome c (Fe^{3+}/Fe^{2+})	+0.260
Cytochrome b_2 (Fe^{3+}/Fe^{2+})	+0.12
Fumarate/Succinate	+0.03
Pyruvate/Lactate	-0.190
$NAD^+/NADH + H^+$	-0.320
H^+/H_2	-0.42
Ferredoxin (Fe^{3+}/Fe^{2+})	-0.43

Such a list places the half-cell reactions in order of decreasing standard _____ , so that the strongest oxidants are at the top, and the strongest reductants at the bottom. This arrangement permits you to predict, given the proper catalysts and standard conditions, that the cytochrome c system will oxidize the cytochrome b_2 system, or that the ferredoxin system will _____ the cytochrome b_2 system. Can we make these predictions quantitative, i.e., calculate at what concentrations of the four reactants the reaction will barely occur spontaneously? _____

44

electrode potential (sometimes called *electrode reduction potential*)

reduce

Yes (This should be apparent, perhaps by reference to the Nernst equation, Item 39.)

45

* An important example will help show you how the values for the free energy change contributes to the understanding of bioenergetics.

Much of the energy obtained by respiration is conserved by the process of oxidative phosphorylation, whereby ATP is synthesized from ADP and P_i. The removal of hydrogen atoms from substrates during catabolism leads often to the reduction of NAD^+ to NADH. We can use the following over-all reaction to show how energy is conserved on the subsequent oxidation of NADH:

$$NADH + H^+ + 3ADP + 3P_i + \frac{1}{2}O_2 \rightarrow$$
$$NAD^+ + 4H_2O + 3ATP.$$

We can, as usual, formally dissect this reaction into an exergonic component,

$$NADH + H^+ + \frac{1}{2}O_2 \rightarrow NAD^+ + H_2O, \Delta G' = -52,400$$
$$\text{cal./mole NADH}$$

and an endergonic component,

$$3\,ADP + 3P_i \rightarrow 3ATP + 3H_2O, \Delta G' = +21,900 \text{ cal.}/$$
$$3 \text{ moles ATP.}$$

For our over-all reaction the free energy change under standard conditions is _____ cal. /mole NADH. Given the appropriate catalysts, is the over-all reaction possible? _____

46

* Under standard conditions, what percent of the energy available from the first reaction can be conserved by coupling to the second to produce the stated over-all reaction? _____ %

Hence we may say that the efficiency of coupling is _____ %, still under standard conditions.

47

* Since the above over-all equation reveals no basis for fusion of the endergonic component into a spontaneous process, this equation must fail to reveal the actual intermediates in oxidative phosphorylation. When NADH is converted into NAD^+, two electrons must be released. These electrons are known to be passed subsequently from one acceptor to another in the electron-transport chain, as illustrated in the accompanying figure.[4]

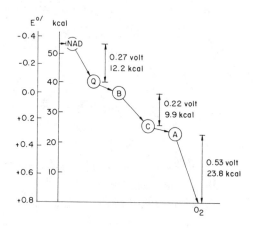

[4] Reproduced with permission from A. L. Lehninger, *Biochemistry*, Worth Publishers, Inc., New York, 1970.

45

−30,500

Yes (since this value is negative)

46

41.8 (21,900/52,400)

41.8

Estimate the free energy change associated with the two steps for which these changes are not inscribed on the figure.[5]

By examining the figure, identify in the list below the steps in electron transport that provide enough energy to allow generation of an ATP molecule.

- [ ] 1. NADH to coenzyme Q
- [ ] 2. Coenzyme Q to cytochrome b
- [ ] 3. Cytochrome b to cytochrome c
- [ ] 4. Cytochrome c to cytochrome a
- [ ] 5. Cytochrome a to oxygen

47

Coenzyme Q to cytochrome b, ca 3.5 kcal.

Cytochrome c to cytochrome a, ca 1.1 kcal.

☑ 1. NADH to coenzyme Q

☑ 3. Cytochrome b to cytochrome c

☑ 5. Cytochrome a to oxygen

48

NADH is generated not only by oxidations occurring in the mitochondrion, but also by such reactions occurring in the cytoplasm and in the smooth endoplasmic reticulum ("microsomes"). One might suppose that the energy conserved by NADH formation would be unable to drive oxidative phosphorylation in those cells where the latter process is confined to the mitochondrion, because the mitochondrion is impermeable to this coenzyme.

There are, however, transport processes in the inner mitochondrial membrane which permit certain sets of analogous anions to be exchanged across the membrane. These exchanges permit reducing equivalents to be transferred between the mitochondrion and the cytoplasm by indirect routes called *shuttles*.

[5] Even if you omitted the items on the oxidation-reduction potential, you will note that knowledge of the free energy changes is quite sufficient to allow you to answer the question.

For example, the cytoplasm contains an NAD-linked glycerophosphate dehydrogenase, and the mitochondrion, an FAD-linked glycerophosphate dehydrogenase. Furthermore transmembrane exchange can occur between dihydroxyacetone phosphate and glycerophosphate. This situation permits reducing equivalents, i.e., electrons, to be imported into the mitochondrion in the form of dihydroxyacetone phosphate/glycerophosphate .

49

There are also important situations where the mitochondrion must transfer reducing equivalents to the cytoplasm. This transfer likewise is not prevented by the impermeability of the organelle to NAD and NADH. In this case the shuttle is believed to involve especially the exchange of mitochondrial malate for cytoplasmic oxaloacetate or an equivalent anion. An NAD-linked malate dehydrogenase is present in each compartment.

The control of such shuttles will undoubtedly prove of fundamental importance to the regulation of metabolism. The net exchange of mitochondrial malate for cytoplasmic oxaloacetate would represent the export of oxidizing/reducing equivalents from the mitochondrion.

50

Some other exergonic oxidation processes take place in the living cell, however, without any apparent conservation of the free energy these reactions can make available. For example, energy conservation apparently does not accompany the action of D-amino acid oxidase or of uricase in the

48

glycerophosphate

49

reducing

"peroxysome" or "microbody" of the hepatocyte. The chemical modification of the substrate probably needs to be seen as the sole function of these processes.

This situation is perhaps not as disturbing as it appears at first. The living organism does not have as its prime function the storage of as much energy as possible, i.e., by the accumulation of as much ATP as possible, or by the generation of gradients as large as possible, or even by the synthesis of as much cellular architecture as possible. Would the conversion of all its NAD^+ to NADH, or all its AMP and ADP to ATP, be compatible with continued life? _____

We can compare a living organism, especially an adult land animal, to a surf-board rider harvesting only enough energy from the surf to maintain his momentum. Competition for energy with other organisms may compel the organism to approach the efficiency of its competitors. Does this competition mean that every process consuming unnecessary energy must be avoided? _____

50

No (NAD^+ and AMP are essential cofactors.)

No (At some stages "wasteful" processes may become necessary to maintain supplies of NAD^+, AMP or ADP.)

REVIEW AND PRACTICE

51

* The conservation of the energy made available by a chemical reaction in the living cell is limited to a form of energy represented by | heat/volume expansion/the intensity with which the reaction tends to go to completion | .

52

* The standard free energy change characteristic of a reaction is equal to | $-RT \ln \frac{[product]}{[reactant]}$ at any given time/ $-RT \ln K_{eq}$ /RTpK | .

53

* Any fully described process showing a positive free energy change is | spontaneous/exergonic/unreal | in the biological context.

54

* The biological strategy to obtain the synthesis of a chemical structure with a higher free energy than its precursors is to _____ .

51

the intensity with which ... completion

52

$-RT \ln K_{eq}$

53

unreal

54

link the reaction
sequentially, or fuse
the reaction with
another process; to
obtain an exergonic
total process (or
similar words; less
specifically, to
couple the process to
an exergonic
reaction)

55

Yes

55

* In chemi-osmotic coupling, we want a chemical reaction to drive an uphill transport, or we want a downhill transport to drive a chemical reaction.

Does a reaction that releases one of its reactants at a point other than the one at which it is received carry the theoretical possibility of producing this effect? _____

56

* Here are shown together a transport process and a chemical reaction. The movement of A from phase 1 to phase 2 is passive, and so is its reverse movement from phase 2 to phase 1.

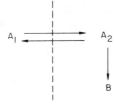

After reaching phase 2, A is converted into another substance, B. If phase 1 is very large, the movement of A_1 into the smaller phase 2 may be caused to continue indefinitely through its depletion by the chemical reaction, which has a high equilibrium constant.

Is an active transport of A produced? _____
Should we regard the interaction seen here as a direct energetic coupling? _____

57

❋ The measurement of oxidation-reduction potentials provides us

☐ 1. unique information about the free energy changes of the oxidation-reduction reactions.

☐ 2. an especially convenient measure of the free energy made available by oxidation-reduction reactions.

58

❋ We can predict whether a given exergonic reaction provides enough energy to be able in theory to produce a given endergonic synthesis from

☐ 1. a determination of the equilibrium constants of the two isolated reactions,

☐ 2. a determination of the standard electrode potentials of the reactions, wherever these measurements are possible.

59

The standard half-cell potential of a given half-cell reaction is obtained by

☐ 1. measuring the potential it yields in isolation.

☐ 2. measuring the overall potential yielded when this half-cell reaction is joined to another half-cell reaction having a known standard electrode potential.

56

No

No (See Items 16–19.)

57

☑ 2. an especially convenient . . .

58

Both responses are correct.

59

☑ 2. measuring the
over-all potential . . .

60

only 3

Yes

61

far from

near to

60

* Beginning from NADH, the electron transport chain of the mitochondrion divides the energy-releasing steps into a series of smaller increments. Among these [only 2/only 3/all] yield enough free energy to cause the synthesis of one ATP molecule. Does such subdivision enhance the possibility of matching the "packets" of energy made available to that required for each energy-conserving synthesis of an ATP molecule?

61

* The following striking paradox is presented by chemical thermodynamics in the biological context:

1. A reaction makes available the largest quantity of energy when it is carried out as [near to/far from] equilibrium as possible.

2. Nevertheless the free energy made available by the reaction is best conserved when the reaction is fused into a more complex process which proceeds over-all as [near to/far from] equilibrium as possible.

Next page 178, Appendix

I. pH AND DISSOCIATION

1. Indicate carefully how you could efficiently prepare 100 ml of $0.1M$ buffers, one with each of the following expected pH values, using the indicated substance plus HCl or NaOH as needed in each case. Assume that the pK' values listed apply under the conditions prevailing.

		pH desired
A.	Acetic acid, $pK' = 4.7$	4.5
B.	Sodium acetate, $pK' = 4.7$	5.1
C.	Glycine, $pK'_1 = 2.3$, $pK'_2 = 9.8$	3.0
D.	Glycine	9.7
E.	Phosphoric acid, $pK'_1 = 2.0$, $pK'_2 = 6.8$, $pK'_3 = 11.8$	7.0

2. Calculate the concentrations of the selected component to be expected in the following solutions:

		pH	calculate
A.	$0.1M$ acetate buffer	4.9	[acetate$^-$]
B.	Acetate buffer; [Acetic acid] $= 0.06M$	4.9	[acetate$^-$]
C.	$0.1M$ glycine buffer	9.5	[$NH_3^+CH_2COO^-$]
D.	A saline solution containing $0.001M$ HPO_4^{2-}	7.1	[$H_2PO_4^-$]

3. In which direction, toward the anode, toward the cathode, or neither, would the following peptides migrate in an electrical field at pH 7.4? At pH 1? A. Alanylglycylserine; B. Methionylhistidylproline; C. Threonylleucylasparagine; D. Phenylalanylvalyllysine

II. KINETICS

1. An enzyme reaction known to follow the Michaelis-Menten equation shows the following initial rates under the conditions shown:

$[S]$ M	Velocity, min^{-1} at $27°$	at $37°$
1×10^{-5}	253	
3×10^{-5}	378	
1×10^{-3}	500	1100

Either by inspection or by plotting, estimate a value each for V_{max} and K_m.

Use the Michaelis-Menten equation to predict the initial rate at $3 \times 10^{-5} M$ and $27°$. Does your result tend to support the correctness of your estimates for V_{max} and K_m?

Suppose we do not know whether this reaction is catalyzed by an enzyme or not. Does the value of $Q_{10}°$ available from the data answer this uncertainty?

2. At what substrate concentration, relative to the K_m, will an enzyme reaction show one-quarter of its maximum rate, according to the Michaelis-Menten equation?

3. Salicylate inhibits the action of glutamic dehydrogenase, whereby α-ketoglutarate and NH_4^+ are formed. Here are shown the initial rates of α-ketoglutarate formation, with and without salicylate present at 40mM. Determine by inspection or by graphing whether or not the inhibition is competitive with the substrate. Also extract a K_m value for glutamate in this enzymatic reaction.

Glutamate concentration (mM)	1.5	2.0	3.0	4.0	8.0	16
mg product/min (without salicylate)	0.21	0.25	0.28	0.33	0.44	0.43
mg product/min (with salicylate)	0.08	0.10	0.12	0.13	0.16	0.18

4. An enzyme composed of four similar subunits shows in its native form a sigmoid form for its plot, v versus $[S]$. After chemical treatment it comes to correspond precisely to the Michaelis-Menten equation. Predict what the chemical treatment has accomplished.

Page 179

5. Summarize the requirements for enzyme assays: Which is to be present in rate-limiting quantity, and which in excess: 1) the substrate; 2) cofactors; 3) an enzyme whose activity is being assayed; 4) an enzyme which is being used simply as a reagent to measure the concentration of a substrate, or the concentration of the product of another enzymatic reaction; 5) the product.

III. BIOENERGETICS

1. Suppose we begin with a mole of a solute dissolved in a liter of water. What free energy change would occur when this solution is diluted at 25° with 9 liters of water? What minimal quantity of work would need to be done (measured in calories) to bring the solution back to its original condition? Assume that activity coefficients equal 1.

3. Calculate the standard free energy changes of the following reactions either from the stated equilibrium constant, or from the standard electrode potential listed above in Item 43:

   A.   Glutamate + oxaloacetate $\rightleftharpoons$ aspartate + α-ketoglutarate, K'_{eq} = 6.8

   B.   Ferredoxin (Fe^{3+}) $\rightleftharpoons$ ferredoxin (Fe^{2+})

   C.   Sucrose + $H_2O \rightleftharpoons$ glucose + fructose, K'_{eq} = 1.33 $\times$ 10^5

Suppose a person consumes 34.2 g (0.1 mole) of sucrose per day. The energy released by its conversion to glucose and fructose would be 700 calories. Is this energy a significant factor in the fattening action of a high sugar intake?

4. When phospho*enol*pyruvate is hydrolyzed the immediate products are ortho-phosphate ion and the enol form of pyruvate:

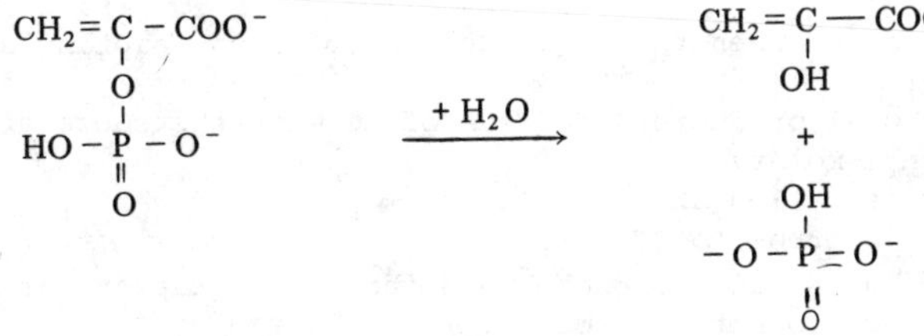

$$CH_2 = C - COO^-$$
$$|$$
$$O$$
$$|$$
$$HO - P - O^-$$
$$\|$$
$$O$$

$$\xrightarrow{+ H_2O}$$

$$CH_2 = C - COO^-$$
$$|$$
$$OH$$
$$+$$
$$OH$$
$$|$$
$$- O - P = O^-$$
$$\|$$
$$O$$

Another reaction promptly follows:

$$CH_2 = C - COO^- \qquad \longrightarrow \qquad CH_3 - C - COO^-$$
$$\quad\quad\quad | \qquad\qquad\qquad\qquad\qquad\qquad\qquad \|$$
$$\quad\quad\quad OH \qquad\qquad\qquad\qquad\qquad\qquad\quad O$$

*enol*pyruvate   *keto*pyruvate

The value of $\Delta G'$ of $-14{,}800$ cal. for phospho*enol*pyruvate hydrolysis covers both of these reactions. Suppose that K'_{eq} for the second reaction is 10^8 at pH 7. Calculate approximately what fraction of the total free energy change listed for the hydrolytic reaction actually arises from the hydrolysis per se, and what fraction from the subsequent rearrangement of the unstable *enol* form. Could we ascribe the highly exergonic character of phospho*enol*-pyruvate hydrolysis to a sequential linkage between these two reactions?

5. The lactate dehydrogenase of heart muscle has an unusually high K_m for the product, pyruvate, compared with the corresponding enzyme in skeletal muscle. This product is an inhibitor of the cardiac enzyme. These features contribute to the circumstance that the enzyme serves in heart muscle mainly to convert lactate to pyruvate, rather than the reverse. Is it possible by "jiggering" the kinetic constants for the substrate and the product to cause the reaction to go in the endergonic direction, e.g., to cause the [pyruvate]/[lactate] ratio to rise higher than its equilibrium value?

6. Complete the following equation for the free energy change produced by a chemical reaction, $Q \rightarrow R$:

$$\Delta G = -RT\ln \frac{[R]eq}{[Q]eq} + RT\ln \underline{\quad\quad}$$
$$\underbrace{\qquad\qquad\qquad} \quad \underbrace{\qquad\qquad}$$
$$(\quad) \qquad\qquad (\quad)$$

For the two terms on the right-hand side of the equation, place letters in the blanks above to indicate which term corresponds to the following descriptions:
A. The value of ΔG when the concentration of each reactant and product is 1 molar.
B. The additional or decreased value of ΔG when the concentrations are other than 1 molar.
C. This term is called ΔG°.
D. This term is equal to $RT\ln K_{eq}$.
E. This term has a value of zero when the concentrations of each reactant and product is 1 molar.

ANSWERS TO TEST PROBLEMS

I. pH AND DISSOCIATION

1. Per 100 ml, use:

   (or mEq)

   A. 10.0 mmoles acetic acid $\underset{\wedge}{\text{(or mEq)}}$

   3.87 mEq NaOH

   B. 10.0 mmoles sodium acetate

   2.85 mEq HCl

   C. 10.0 mmoles glycine

   1.66 mEq HCl

   D. 10.0 mmoles glycine

   4.43 mEq NaOH

   E. 10.0 mmoles $H_3 PO_4$

   16.1 mEq NaOH

2. A. [Acetate$^-$] = 0.061 M

   B. [Acetate$^-$] = 0.095 M

   C. [$H_3^+ NCH_2 COO^-$] = 0.067 M

   D. [$H_2 PO_4^-$] = 0.0005 M

3.

		pH 7.4	pH 1
A.	Alanylglycylserine	neither	cathode
B.	Methionylhistidylproline	cathode	cathode*
C.	Threonylleucylasparagine	neither	cathode
D.	Phenylalanylvalyllysine	cathode	cathode*

*more rapidly at this pH

II. KINETICS

1. The first two observations can be seen to lie in the central (B) portion of the Michaelis-Menten curve. Because the third concentration is so much higher, it must produce a nearly maximal velocity. Accordingly V_{max} is about 500 min^{-1}. Since 1×10^{-5} M produced close to half that rate, it approximates the K_m value. The velocity calculated for $3 \times 10^{-5} M$ using these parameters, $\dfrac{500 \times 3}{1 + 3}$ = 375 min^{-1}, falls very close to the rate observed at $3 \times 10^{-5} M$, to confirm the values for K_m and V_{max}. The Q_{10° of 2.2 seen between 27° and 37° does not help us decide whether the reaction is enzymatic or not, since chemical reactions in general have Q_{10° values of about 2, whether enzyme-catalyzed or not.

APPENDIX

2. The problem restated mathematically reads: At what concentration does

$$v = \frac{V_{max}}{4}? \text{ Since } \frac{V_{max}}{4} = \frac{V_{max} \cdot \frac{1}{3} K_m}{1\frac{1}{3} K_m}, \text{ a concentration of } \frac{1}{3} K_m \text{ will produce a quarter}$$

maximal velocity.

3. Inspection will show that the effect of the inhibitor is not well overcome by increasing the substrate concentration; hence the inhibition is not of the competitive type. A Lineweaver-Burk plot shows that K_m for glutamate was unchanged by the inhibitor; hence the inhibition is specifically of the non-competitive type. K_m = ca 2.25 mM.

4. The chemical treatment appears to have eliminated the interactions seen among the four subunits in the native enzyme, whereby the filling of one site intensifies binding by other sites. (A change in conformation, a loss of conformational instability, or perhaps even a dissociation into the subunits, could account for this change.)

5. The substrate must be in excess when the assay is made for enzyme activity, but rate-limiting when the assay is for the amount of substrate present. Cofactors should be present in excess. The enzyme should be present in rate-limiting quantity when its activity is being assayed; it should be present in excess when it serves as an analytical reagent. Products should be as nearly absent as possible.

III. BIOENERGETICS

1. $\Delta G = 2.303 \text{ RT } \log \frac{0.1}{1} = -1364 \log 10 = -1364$ cal./mole. With maximal efficiency, work equivalent to an equal number of calories would be required to reconcentrate the solute to one molal.

3. A. $\Delta G' = -1140$ cal./mole; B. $\Delta G' = -n\,FE = -23,061 \times (-0.43) = 9916$ cal./mole; C. $\Delta G' = -7000$ cal./mole. The energy released by the hydrolysis of the disaccharide, 700 cal., is insignificant compared with the 136,800 calories available from its catabolism. Note that the kilocalorie or Cal. is usually used in dietetic and nutritional calculations. Furthermore, this energy is released by intestinal digestion, and is conserved only to the extent that the absorption of the monosaccharides might conceivably require less energy because intestinal digestion can release them at high concentrations.

4. $\Delta G'$ for the *enol-keto* rearrangement = -2.303 RT log 10^8 = $-10,912$ cal./mole. $\Delta G'$ for the hydrolytic step alone = $-14,800 - (-10,912) = -3,888$ cal./mole. About 26% of the total free energy change can be attributed to the hydrolytic step *per se*, and 74% to the consequent rearrangement. Hence the sequential linkage of the hydrolysis to the rearrangement leads to the highly exergonic total effect.

5. Manipulation of the affinities of an enzyme for its substrate and its product cannot cause the reaction, fully described, to proceed in the endergonic direction. The same is true for transport; manipulation of the affinities of the carrier for the transported substrate at the two surfaces of a membrane cannot produce uphill transport except at the usual cost in external energy. In ordinary biological system this energy must be provided by an energetic coupling that yields an overall process which is exergonic and therefore spontaneous.

6.
$$\Delta G = \underbrace{-RT\ln \frac{[R]_{eq}}{[Q]_{eq}}}_{(A,\ C,\ D)} + \underbrace{RT\ln \frac{[R]_{actual}}{[Q]_{actual}}}_{(B,\ E)}$$